Services and Business Process Reengineering

The book series aims at bringing together valuable and novel scientific contributions that address the critical issues of software services and business processes reengineering, providing innovative ideas, methodologies, technologies and platforms that have an impact in this diverse and fast-changing research community in academia and industry.

The areas to be covered are

- Service Design
- Deployment of Services on Cloud and Edge Computing Platform
- Web Services
- IoT Services
- Requirements Engineering for Software Services
- Privacy in Software Services
- Business Process Management
- Business Process Redesign
- Software Design and Process Autonomy
- Security as a Service
- IoT Services and Privacy
- Business Analytics and Autonomic Software Management
- Service Reengineering
- Business Applications and Service Planning
- Policy Based Software Development
- Software Analysis and Verification
- Enterprise Architecture

The series serves as a qualified repository for collecting and promoting state-of-the art research trends in the broad area of software services and business processes reengineering in the context of enterprise scenarios. The series will include monographs, edited volumes and selected proceedings.

More information about this series at http://www.springer.com/series/16135

Roberto Senigaglia · Claudia Irti ·
Alessandro Bernes
Editors

Privacy and Data Protection in Software Services

Springer

Editors
Roberto Senigaglia
Department of Economics
Ca' Foscari University of Venice
Venice, Italy

Claudia Irti
Department of Economics
Ca' Foscari University of Venice
Venice, Italy

Alessandro Bernes
Department of Economics
Ca' Foscari University of Venice
Venice, Italy

ISSN 2524-5503 ISSN 2524-5511 (electronic)
Services and Business Process Reengineering
ISBN 978-981-16-3048-4 ISBN 978-981-16-3049-1 (eBook)
https://doi.org/10.1007/978-981-16-3049-1

This Springer imprint is published by the registered company Springer Nature Singapore Pte Ltd.
The registered company address is: 152 Beach Road, #21-01/04 Gateway East, Singapore 189721, Singapore

Foreword

Relations (between persons, between entities and between contexts) are nowadays increasingly digital; they develop online, in a context characterised by the hegemony, now not only commercial, of online platforms. The social, cultural, legal and anthropological revolution brought about by digital technology has taken place at such a speed that the law has been forced to take up an unequal stride. The regulatory sector that has been able to provide an adequate and far-sighted regulation (because it is based on the combination of solid principles and some essential rules) is that of data protection. This discipline is particularly relevant for the governance of the digital world as it acts on the conditions of circulation of what, like data, constitutes the primary factor of the digital ecosystem, the object of a fundamental right and, at the same time, an increasingly attractive economic resource for platform capitalism.

This demonstrates how, in a subject such as this—which is exposed to an incessant evolution both in terms of interpretation and application and to the continuous comparison of the rule with concrete cases which are always new and with unprecedented characteristics—the interpreter has an essential need for guides, tools capable of guiding the analysis and the choice among the options which are often possible, among the various meanings which the same provision can support. Hence, the scholarly reflection is irreplaceable on a subject that has recently known—alongside several and important innovations of the jurisprudential formant—the 'revolution' of the GDPR and the relative national regulations of adaptation, as well as the very important novelties of the Directive 2016/680 (so-called Law Enforcement Directive), with the internal transposition law.

In such a complex regulatory framework, it is therefore extremely essential to adopt—as this book does—a synoptic perspective, which constantly refers to every single provision within the broader context in which it is placed. Above all, the ability to grasp the social function of this extraordinary right—which has never been a 'tyrant'—in the balance with other relevant legal interests, recalling that the processing—like the GDPR states—should be designed to serve mankind.

This book illustrates important aspects of the data protection discipline, something profoundly innovated, first and foremost, by the choice of the European legislator to replace a harmonisation instrument such as the directive with an instrument such as the regulation, which is also applicable to the processing of personal data by a controller or processor not established in the EU, where the processing activities are related to an offer of goods or services to data subjects in Europe or the monitoring of their behaviour.

All this is being pursued in a framework that constitutes a new challenge for all operators involved, starting with the controller. The latter is given new possibilities to use data, in the sharing economy, with a much more dynamic and articulated relationship between consumer and business. Equally significant, as underlined in several passages of the book, is the strengthening of the rights of the data subject, with new implications such as data portability—which makes it possible to reassemble the pieces of the mosaic of our digital self, while also protecting competition from lock-in phenomena—and the right to erasure.

The rights-oriented approach on which the GDPR is based also implies a shift from mainly remedial protection, i.e. at a later stage, to essentially preventive protection, with a simultaneous strengthening of the rights of the data subject. This protection is particularly relevant because of the risk that traditional safeguards, such as consent and information, become relative in the context of the IoT (as it is well underlined) and of massive data collections, which are often beyond individual control due to the fragmentation of the data management process along a chain with multiple links.

But perhaps the most significant expression of the preventive approach concerns the overall responsibilities of the controller, towards the adoption of a business strategy founded on data protection. This also considers that the same violation of the principle of accountability integrates, like the non-compliance with the other principles, the extremes of an autonomous infringement. On this aspect, the book offers some points of notable interest and particular help for the interpreter, which will be precious also in the applicative phase, plumbing the various declinations of data protection in the most varied sectors: from the app economy to the power of online platforms, from the protection of minors to the IoT and from the relationship between privacy and public health to cybersecurity.

It also correctly underlines the European dimension in which data protection lives, as a result of the Court of Justice's case law and of the importance of cooperation procedures, which are perhaps the most eloquent expressions of the European aspiration to achieve 'one continent, one law'.

The distinctive feature of the book, which makes it even more appreciable, is its integration, with continuous cross-references, of the 'operational' and concrete dimension with the theoretical perspective and the normative analysis, enriched by the comparison with the jurisprudential solutions that have mostly characterised the data protection discipline and its principles in the living law.

In this sense, for example, the reflections on location data and tracing apps are valuable, especially in light of the development they have had for epidemiological purposes during the pandemic.

Also important are the reflections on AI-based business decisions and blockchain, which will increasingly represent a fundamental component of our future.

The contribution on personal data as a counter-performance is also very important, as it represents a challenging issue on which Europe can really provide a guide to prevent the mere logic of profit from prevailing over the protection of fundamental rights.

Additionally, the reflection on the legal effects of the digital divide, which is today the modern version of inequality and represents the area in which truly social democratic public policies should invest their best resources, is important.

The data protection regulation is, therefore, a great step forward in the direction of balanced governance of the technological innovations that have profoundly changed our society. However, the success of this 'gamble' will depend on its social stability, on its ability to become the form and rule of action for citizens and private and public entities. Papers like this one are an important step in that direction.

Solopaca, Italy

Pasquale Stanzione
President of the Italian Data Protection Authority

About This Book

In the world of today, it has started to become difficult to distinguish a data-driven economy from the social system as a whole, as well as the real-life of a natural person from its virtuality. Individual interests, choices and data have already been used by software services not only for commercial purposes across the digital landscape but also for helping users, consumers and citizens to obtain digital products and to access daily services more easily, as well as making people stay connected on social platforms.

Nevertheless, personal data have been considered the 'new oil' of present times. Huge amounts of data are increasingly accumulated by Tech Giants—and public governments—in the globalised world. Digital companies are currently able to analyse (Big) data in order to anticipate granular decisions and target end-users with personalised advertisements. The latter task belongs not to human beings, but it is carried out by Artificial Intelligence means, algorithms and machine learning, which are designed for executing thousands of operations in one second. In this respect, data mining and personal information flow have given rise to a new era, better known as the age of information economy, whose degenerations, in terms of nudging individuals, lead to 'surveillance capitalism'.

A concrete example of the above-mentioned portrait can be seen in the reengineering process of standard software coding and development in the field of mobile applications (so-called 'apps'). Apps seek to appear more user-friendly for their users, up until the point that they are capable of clinching an agreement simply with a few touches. The associated business model involves not only the e-commerce sector but can be seen as a standard of many other information society services, in which individuals actually develop their own personality and have social interactions, for obtaining ever-greater data from their users (e.g. social networks).

All those solutions usually rely on digital identities and user accounts, which are based upon 'information relating to an identified or identifiable natural person', under the Regulation (EU) 2016/679—General Data Protection Regulation (GDPR)—which defines the concept of personal data.

Efficiency and wealth should not have the upper hand over individuals' fundamental rights and freedoms, taking into particular account human dignity. In this direction, the EU law approach is moving towards a proactive management strategy.

In essence, it is strongly required to make previous risk assessments and harsh controls for better evaluating the forthcoming data usage in online systems and web apps. The data controller is asked to implement appropriate technical and organisational measures, as well as security tools for the processing of personal data, which have to be designed to better enforce data minimisation, cybersecurity, system resilience, etc., following the principle of transparency and the multiple challenges led by new technologies (e.g. blockchain). Accordingly, a renovated approach concerning technical means to be adopted must be carried out by developing privacy-friendly and trusted software products from the very beginning, in accordance with data protection by design and by default settings.

Building up such a pre-set data management system is still far from increasing citizens' awareness on which data are collected, for what purposes and so on, since the data flow is far from being at least potentially understood by data subjects, further enabling them to fully exercise their rights (right to data access, right to erasure, right to data portability, etc.).

In the age of datafication, any breach of the right to privacy and data protection is perceived by individuals as less relevant than other restrictions of personal freedoms. For example, when the trader supplies digital contents or digital services, the consumer usually pays more attention to the price than to personal data—not strictly functional for the performance of a contract—provided to the trader (although the respect of fundamental values, such as personal identity, wishes to not consider data as a mere commodity). Besides, cookies, which also serve to track online users' behaviour, are often considered annoying pop-up ads which have to be closed immediately. Most of these aspects intensify the problem, such as in the case of children who have easy access to the Net and who settle everyday transactions without adult supervision.

Following this heuristic perspective, this book is structured in three sections: 'Problems' (Part I), 'Perspectives' (Part II) and 'Applicable Solutions' (Part III). Part I identifies the current discussions about the issues that the regulation of Information and Communication Technologies (ICT) poses to the dominant datafication process, especially for the processing of personal data carried out by digital service providers. Part II identifies several points of view that each actor who deals with data protection in the online environment has to face not only in the abstract but also in the context, by covering the whole software development life-cycle. In Part III, a special focus is given to the use of certain technologies, which may certainly lead to enormous benefits for both end-users and digital companies, if correctly set up from both the legal and technical sides.

The passage, from a formalistic vision and a defensive approach of information privacy to a factual approach and subsequent risk assessment, surely requires a change of paradigm. It needs to experience a new operational approach amid technological development and individuals' fundamental rights and freedoms, by placing human beings at the centre of the discourse. Accordingly, the aim of this book is to create a bridge between two 'lands' that are usually kept separate: technical tools and legal rules, instead, they should be bound together for moulding a special 'toolbox' to solve present and future issues. The following pages are intended to contribute to this 'toolbox' in the international landscape: not only in the area of legal studies, but they also address how to make technology and law work closely with engineers' and computer scientists' fields of expertise, who are increasingly involved in tangled choices on daily programming and software development pathways.

Roberto Senigaglia
Claudia Irti
Alessandro Bernes

Contents

Applicable Solutions

Editors and Contributors

About the Editors

Roberto Senigaglia is full professor of Private Law at Ca' Foscari University of Venice (Italy), where he teaches Private Law, Family Law, Juvenile Civil Law and Fundamental Rights and Privacy. He has Ph.D. in European Civil and Commercial Contracts Law. His scientific activity is focused on issues in the areas of Personal and Family Law and Law of Obligations and Contracts. His research focuses, in particular, on the impact of Fundamental Rights in Private Law relationships. He authored numerous contributions, including monographic ones, in the different areas of Private Law.

Claudia Irti is associate professor of Private Law at Ca' Foscari University of Venice (Italy), where she teaches Private Law, Family Law and Tourism Law. She holds Ph.D. in Comparative Civil Law from the University of Firenze (Italy) and was graduated magna cum laude from the University of Bologna (Italy). She has been a visiting scholar at the University of Berkeley (California, USA), at the University of Regensburg (Germany) and at the Max Plank Institute of Hamburg (Germany). She published numerous articles, book chapters and one monograph on several Private Law and Family Law matters.

Alessandro Bernes (Trieste, October 1990) is assistant professor of Private Law at Ca' Foscari University of Venice (Italy), where he teaches Data Protection Law and Tourism Law. After his master's degree in Law (University of Trieste), he obtained Ph.D. in Law, Market and Person at Ca' Foscari University. His postdoctoral studies focus on technical and organisational measures needed to ensure better personal data processing in accordance with scientific research purposes. He has published articles and essays in disparate scientific academic journals on the subject of personal data protection, real estate law and antitrust damage claims.

Contributors

Alessandro Bernes Ca' Foscari University of Venice, Venice, Italy

Fabio Bravo University of Bologna, Bologna, Italy

Carmelita Camardi Ca' Foscari University of Venice, Venice, Italy

Alberto De Franceschi University of Ferrara, Ferrara, Italy

Giusella Finocchiaro University of Bologna, Bologna, Italy

Alberto Maria Gambino European University of Rome, Rome, Italy

Andrea Maria Garofalo Ca' Foscari University of Venice, Venice, Italy

Claudia Irti Ca' Foscari University of Venice, Venice, Italy

Sara Landini University of Florence, Florence, Italy

Alessandro Mantelero Polytechnic University of Turin, Turin, Italy

Barbara Pasa Università IUAV di Venezia, Venice, Italy

Carolina Perlingieri University of Naples «Federico II», Naples, Italy

Dianora Poletti University of Pisa, Pisa, Italy

Giorgio Resta Roma Tre University, Rome, Italy

Vincenzo Ricciuto Tor Vergata University of Rome, Rome, Italy

Roberto Senigaglia Ca' Foscari University of Venice, Venice, Italy

Camilla Tabarrini Ca' Foscari University Venice, Venice, Italy

Davide Tuzzolino European University of Rome, Rome, Italy

Giuseppe Vaciago University of Insubria, Varese, Italy

Giuliano Zanchi Ca' Foscari University, Venice, Italy

Vincenzo Zeno-Zencovich Roma Tre University, Rome, Italy

Problems

Transparency of Digital Providers and Digital Divide

Giusella Finocchiaro

1 Towards the Consolidation of the Single European Market

Since 2015, the European lawmaker has been committed to the creation of a European Digital Society, in which the free movement of goods, people, services and capital is guaranteed.

The aim is to exploit the potential of new information and communication technologies in order to improve access to digital goods and services throughout Europe and to create a favourable environment and a level playing field for the development of digital networks and innovative services.

As part of this process, which is still young but already well-defined, the European lawmaker has succeeded in achieving some important results which, taken together, make it possible to tackle the fragmentation of the market, develop the necessary digital infrastructures and promote the digitization of European industry. Some illustrative examples are represented by the reform of the personal data protection legislation; the establishment of the regulatory framework to ensure cross-border portability of online contents; and, finally, the agreement to unlock the market for electronic commerce by eliminating unjustified geographical blocks. All these were deemed to be necessary measures to remove the technical, legal and bureaucratic barriers that hindered technological development and cross-border relations, including those online.

This has been partly possible also thanks to an appropriate choice of the instruments of harmonization and standardization of the law. Indeed, in order to remove the legal obstacles created by the heterogeneity of national legislations, the European lawmaker often used regulations which, as is known, are directly applicable in all

G. Finocchiaro (✉)
University of Bologna, Bologna, Italy
e-mail: giusella.finocchiaro@unibo.it

R. Senigaglia et al. (eds.), *Privacy and Data Protection in Software Services*,
Services and Business Process Reengineering,
https://doi.org/10.1007/978-981-16-3049-1_1

Member States of the European Union, without requiring national transpositions. It is the use of a standardization instrument of the law for the European States, rather than of a harmonization one (such as the directive), which allowed to cancel out those small differences which make it extremely difficult to fully achieve a single market.

Hence, it is not by chance that two main pillars of the European Digital Strategy have been prepared using the legal form of the regulation. These are the Regulation (EU) 2016/679 on the protection of personal data (Finocchiaro 2019) and the Regulation (EU) 910/2014 on electronic identification and trust services for electronic transactions in the internal market (Delfini and Finocchiaro 2017). The first one, by repealing the previous Directive 95/46/EC, has innovated the personal data protection legislation, focusing—from the title—on the free movement of data and their protection. The second has dealt with the problem of online identification, by also regulating electronic signatures and the so-called trust services.

In both cases, the legislative intervention was motivated by the concern for the largely fragmented landscape of the national rules on the protection of personal data and on electronic identification, respectively, and by the widespread legal uncertainty concerning the application of the rules at the Union level. The need was therefore to ensure a uniform implementation of the legislation, in order to foster a climate of trust conducive to economic growth, especially in online environments.

Accordingly, the two Regulations, considered from a unified perspective, clearly indicate the purpose of the European lawmaker to shape a Digital Single Market (DSM) (De Franceschi 2016), as well as to strengthen the European position in the global competition, affirming a homogeneous approach based on the principles of the Charter of Fundamental Rights of the European Union. In particular, the European lawmaker decided to reinforce trust in electronic transactions within the internal market by ensuring the protection of personal data and the security of online relationships, thereby also increasing the effectiveness of public and private online services in the European Union.

These two fundamental pillars of the European strategy are now supplemented by the Regulation (EU) 2019/1150 which promotes fairness and transparency rules for online intermediation services to protect business users and the draft Regulation for a single market for digital services, better known as the Digital Services Act (DSA). This latter legislative initiative, in particular, has become necessary due to the global economic and social transformation brought about by the growth of new information society services which have changed the way citizens and businesses communicate, consume and establish personal and commercial relationships. On the one hand, the development of digital services, such as online platforms, has made the internal market more efficient, fostering innovation and facilitating trade and access to new markets. On the other hand, this potential has not been supported by a concurrent regulatory adjustment that responds to the challenges and risks associated with the development of this new dimension, concerning the protection of users, the responsibility of the providers and the access to the market.

After twenty years since the adoption of Directive 2000/31/EC on electronic commerce (e-Commerce Directive), it is now clear that this instrument is no longer adequate to regulate an increasingly digitized world. In the current historical moment, then, the health emergency caused by the spread of COVID-19 has shown even more dramatically the absolute relevance of digital technologies and the dependence of our economy on digital services.

2 A New Liability Regime: From the Model of Dir. 2000/31/CE to Date

By amending the e-Commerce Directive, the DSA aims to establish a framework of accountability and rules of due diligence and transparency for providers of intermediary services, including social media and marketplaces, to ensure a safe and trusted environment for users. In particular, the draft Regulation aims to improve the online safety of users throughout the Union and to ensure the protection of their fundamental rights by introducing clear due diligence obligations, including notice-and-action procedures to report and remove illegal contents. Likewise, the obligation for some online platforms to request, verify, store and publish information about traders using their services is intended to guarantee a more trustworthy and transparent online environment for consumers.

Therefore, it is clear the radical change of perspective, not only legislative but also historical and functional, made by the European lawmaker with respect to the liability exemption regime originally provided for by the e-Commerce Directive.

As it is well known, the framework laid down in the e-Commerce Directive is mainly characterized by the adoption of a model of limited liability, or even of liability exemption, for digital service providers. Instead of providing a general liability regime for online intermediaries, the Directive, through the so-called 'safe harbour' principle, establishes specific rules according to which these subjects are exempt from liability under certain conditions related to the assumption of the provider's neutral and passive role. In addition to this framework of conditional exemptions, the Directive does not foresee a general obligation to monitor the online contents of the users: in other words, it is prohibited to require online service providers to actively seek facts or circumstances indicating an illegal activity or contents uploaded or transmitted through their services.

The European lawmaker's leanings for a highly flexible regime without clear-cut obligations were originally due to the awareness of the historical context in which the Directive was approved. Back then, the circumstances and goals of the lawmaker were different. At the beginning of the new millennium, the web was in its infancy and wide-ranging legislation was needed to promote the development of the digital market. Imposing obligations and burdens, including economic ones, on service providers, would have discouraged the development of digital services and online transactions, to the detriment of web users themselves to whom these burdens inevitably would have been passed on. Therefore, although a legal framework to regulate the web was necessary, the development of digital services and online relationships could not be undermined by it. For this reason, the European lawmaker opted for flexible legislation, which included 'non-liability' rules rather than strict rules followed by significant sanctions in case of non-compliance. Of course, this is only one of the possible interpretations that can be formulated, but it is probably the most accredited from a functional and economic point of view.

In any case, much has changed over the last twenty years, not only in the way relationships—both those between people and those having economic relevance—are carried out in the digital world but also in society's perception of the digital dimension. In other words, if twenty years ago, when we were talking about digital, we could still clearly distinguish between, as often referred to, the so-called 'real' and the so-called 'virtual', today this distinction has become increasingly blurred and the digital is considered one of the dimensions in which human action is expressed, also from a legal point of view. With regard to this perception, also public opinion has changed, and many web users wonder why, in some cases, there is no liability for the provider. While this question was probably not even asked twenty years ago, today it is often put forward and needs to be answered.

Furthermore, the rules laid down in the e-Commerce Directive have been progressively eroded by the case law and by the lawmaker itself in response to the demand for providers' accountability that has become increasingly pressing.

As for the case law, judges have, actually, become the framers of the law, creating liabilities even where the e-Commerce Directive did not identify them.

Indeed, many Italian judgments have charged the providers with responsibility, like the historic ruling known as *Google v. Vivi Down*. In the grounds of the first instance judgment, the Court of Milan stated that "there is no codified legal obligation, at least until today, that requires internet service providers to carry out a prior control of the innumerable series of data that pass every second in the meshes of the managers or owners of the websites (…). But, on the other hand, there is also no 'boundless internet prairie' where everything is allowed and nothing can be forbidden, on pain of the worldwide expulsion of the people of the web" [freely translated by the Author]. The present case concerned the dissemination on Google Video, through the AdWords service, of a video that filmed bullying committed by a group of young men against a minor affected by Down's syndrome. After assessing the qualification of Google as hosting or as a content provider, the judges denied the existence of an obligation for the society to monitor in advance the content transmitted on the network at the request of users. As a result, the Court of Milan ruled out the possibility of convicting the provider for not having prevented the crime committed by the user who uploaded the video. However, given the profit that the provider drew from the service it offered and the overall methods of providing the service itself, which allowed him "to manage, index, organize the data contained in the video uploaded on the platform", the judges held the leaders of Google's Italian division criminally liable at least for the purposes of the Italian Data Protection Code, and thus in relation to the offence of unlawful processing of personal data.

This logical-argumentative process has been decisively characteristic of the Italian case law and, in the last phase, also of the European one.

In this regard, it might be useful to recall the case law of the Court of Justice of the European Union (CJEU). While this case law confirms the prohibition of a general preventive control mechanism, two aspects of the CJEU's rulings deserve to be specified. On the one hand, the CJEU does not exclude the possibility for national judges to impose filtering obligations on content uploaded or transmitted online. On the other hand, the CJEU states that, after an injunction from the judicial authority, providers are required to block their users' access to a website containing materials in violation of copyright, as well as to search for and delete defamatory contents.

Against this background, however, one should also consider the ruling of the Italian Court of Cassation of 19 March 2019, nos. 7708 and 7709. On this occasion, while acknowledging the lack of an obligation of surveillance and early, general, and permanent activation on the provider, the judges found the existence of an obligation to remove the offences of which they have become aware. In other words, the judges imposed a duty of qualified diligence on the providers so that the latter become legally responsible for the offence perpetrated through their services when they can be accused of inaction in preventing its continuation.

Therefore, the courts interpreted a social need that called for a safe centre of attribution of responsibility to be identified in the provider through the construction of alternative and new rules, to mitigate the regime on 'non-responsibility' dictated by the e-Commerce Directive.

As for the legislative intervention, the European lawmaker has contributed to the erosion of the principle of conditional exemption imposed by the e-Commerce Directive, by intervening in the field of copyright with the recent Directive (EU) 2019/790 of 17 April 2019 on copyright and the related rights in the DSM. In particular, with this Directive, the European lawmaker has introduced a double-track liability mechanism for providers of online content-sharing services that ensure access to copyright-protected content and for other protected materials uploaded by their users.

Indeed, in addition to expressly provide for the disapplication of the limitation of liability established under the e-Commerce Directive, the Directive 2019/790 requires these providers, first of all, to obtain the authorization of the right-holders of the works that are communicated or made available to the public. Where no authorization is granted to service providers, the Directive 2019/790 requires them to be responsible for any unauthorized act of communication, if they have not or cannot demonstrate that they have made the best efforts, in accordance with high standards of professional industry diligence, to prevent unauthorized works and other subject-matters from being available in their services.

The European lawmaker has, actually, raised the level of diligence required of providers, imposing greater accountability and demonstrating to know the modern techniques of interception of uploaded content that make it possible to ask for greater attention.

In the wake of the e-Commerce Directive, then, the European lawmaker takes up the concepts of notice-and-take-down, further specifying that providers are responsible for unauthorized acts of communication to the public of protected works or other subject-matters where, after receiving a sufficiently reasoned report, they do not act promptly to disable access to or to remove from their websites the works and the other subject-matters flagged in the report.

The European lawmaker made it clear that this provision does not result in the imposition of a general obligation to monitor, thus remaining faithful to the legal tradition laid down in the e-Commerce Directive. However, as a matter of fact, the European lawmaker has created a system of rules to make the providers of online content-sharing services responsible for copyright-protected content. These subjects are now held responsible for the unauthorized publication of contents both in the event of negligence in monitoring and in the event of inactivity following notification of the offence committed through their services.

3 Transparency in the Digital Providers' Activity

In the model of liability that has been emerging in recent times, also thanks to the intervention of the courts, a further parameter of assessment of the behaviour of the digital service providers become central, with reference to how the relations with other economic operators and web users take place.

In this multifaceted context, it is necessary to frame the issue of transparency, which is not only related, as it is usually thought, to the protection of personal data and of the data subjects but also related to the behaviour of digital providers towards users.

Regardless of how the concept of transparency is specifically implemented in each sector and legislative *corpus*, as it will be further explained below, the lawmaker considers transparency to be a key element common to the whole European Digital Strategy in order to ensure a consistent level of protection throughout the Union and to prevent and eliminate disparities and legal uncertainty that can hamper the free movement of personal data and the creation of a DSM.

3.1 Transparency in the GDPR

Within the context of the aforementioned Regulation (EU) 2016/679 on the protection of personal data (better known as the General Data Protection Regulation—GDPR), transparency is one of the fundamental principles governing the processing of personal data (Finocchiaro 2019).

Indeed, pursuant to Article 5(1)(a) GDPR, the data controller, namely, who determines the purposes and means of the processing of personal data, is required to process personal data in a lawful, fair and transparent manner towards the data subjects. Therefore, the European lawmaker refers to the more general principle of good faith, which includes transparency in the behaviour of the data controller.

Compliance with this principle, which is actually a rule of conduct of the data controller, produces a very precise result for the data subjects. Indeed, the latter acquire an important tool of power and control over their data, since, as a consequence of the obligation of transparency of the data controllers, they must be informed of the existence, purposes and methods of the processing of personal data.

The principle of transparency is thus inevitably linked to the very essence of the right to protection of personal data, consisting in the right to have control over these data: the recognition of this right, as well as its practical implementation, is indeed possible only if there is a general obligation to provide information to the data subjects.

The communication of information is therefore crucial for the data subjects to be fully aware of what happens to their personal data and to learn the essential elements for the exercise of their rights. However, in order for this tool of power to be effective, information intended for the data subjects must be concise, easily accessible and intelligible, by using plain and clear language and, where appropriate, standardized icons.

Consequently, the principle of transparency is also a parameter for assessing the responsibility of the data controllers who must prove that they have correctly fulfilled their transparency obligations. To this end, the GDPR promotes the establishment of data protection certification mechanisms and data protection seals and marks which enhance transparency, allowing, on the one hand, the data controllers to demonstrate their compliance with the GDPR and, on the other hand, the data subjects to quickly assess the level of data protection.

3.2 The Transparency of the Digital Services Act

As mentioned before, the DSA identifies a number of due diligence and transparency obligations for digital service providers, which are characterized by the nature of the service and the size of the provider. This differentiated regime allows burdening only the providers who, for the nature and the size of their own activity, can potentially represent a risk to web users.

In particular, the DSA establishes basic obligations applicable to all providers of intermediary services and additional obligations exclusively for providers of hosting services and, more specifically, for online platforms and very large online platforms.

Among the most relevant in terms of transparency, one should recall the obligation, for all providers, to establish within the terms and conditions of the contract the restrictions they impose on the use of their services and to act responsibly in the implementation of these restrictions (Art. 12), as well as the obligation to communicate periodically on the activity of moderation of contents that are illegal or contrary to the terms and conditions laid down by the providers (Art.13).

Moreover, for hosting providers, two further obligations that touch upon the issue of transparency are established. Indeed, providers of hosting services are required to adopt easily accessible and user-friendly notice-and-action mechanisms aimed at allowing third parties to report the presence of alleged illegal content (Art. 14). In addition, if they decide to remove or disable access to specific content, that decision must be justified and accompanied by a statement of reasons that shall include the information specifically indicated in Article 15. The *rationale* behind this tightening of obligations for hosting providers is to be found in the role they can play in the fight against illegal online content. Indeed, since their activity mainly consists of the large-scale storage of information provided by users autonomously or at the request of the service recipients, the hosting providers can have access to an extensive number of contents.

Finally, online platforms, especially very large ones, namely, those offering services to a number of average monthly active recipients in the Union equal to or higher than 45 million, are subject to additional obligations intended, on the one hand, to enhance transparency and, on the other hand, to ensure the management of so-called systemic risks. As for the first purpose, it is worth emphasizing that the online platforms must implement an internal complaint-handling system in relation to decisions on contents allegedly illegal or incompatible with the platforms' terms and conditions (Art. 17); they must cooperate with certified out-of-court dispute resolution bodies to resolve any dispute with the users of their services (Art. 18); they must give priority to notifications of illegal contents submitted by entities which have awarded the status of trusted flagger (Art. 19); they must implement the prescribed measures against misuse (Art. 20); they must inform the competent judicial authorities in the event that they become aware of information which gives rise to suspicions of serious criminal offences involving a threat to the life or safety of persons (Art. 21); they must ensure the reliability and traceability of traders who use their services to conclude distance contracts with consumers (Art. 22); they must publish periodic reports on their activities of removing and disabling illegal contents or contrary to their terms and conditions (Art. 23); they must comply with transparency obligations in relation to online advertising (Art. 24). As to the very large online platforms, for the limited aims of this paper, it is sufficient to point out that the DSA provides for specific and additional obligations on transparency relating to the periodic reporting on their activities (Art. 33).

4 The Web User as a Consumer

The above brief overview of the obligations placed on digital service providers clearly shows that there is a system of rules intended to protect the web users who, in addition to enjoying a wide range of information on the use of digital services, have at their disposal the tools of reporting, complaining, out-of-court appealing to enforce their rights.

Compared to the e-Commerce Directive, it is worth noting a significant change of paradigm, whereby now, as for the protection of personal data and consumer protection regulations, an attempt is being made to strengthen the position of the individual user (Colangelo and Maggiolino 2019). While in the e-Commerce Directive the service provider was in a privileged position where they enjoyed a regime of limited liability, now it is the web user to be at the heart of the protection afforded by the draft Regulation.

Underlying this change is the awareness of the European lawmaker of the new risks arising from digitization and linked, in particular, to the development of new technologies that can predict the users' behaviour and the involvement of big players—the so-called Tech Giants—able to influence the preferences and choices of users. While the digital market can be beneficial in some respects by offering a wider range of services and products, often at competitive prices and without the obstacle of physical borders, it also conceals pitfalls for less savvy users, who may find themselves making choices in an uninformed and not wholly free manner. Profiling, targeted advertising and big data analytics systems are paradigmatic to understand how and to what extent digitalization is able to reduce the self-determination and contractual autonomy of users, whose will is increasingly anticipated and conditioned.

In this context, attention should be paid also to the increase in the level of information and technological asymmetry between the operators of the web and their users (Hoffman 2016). Indeed, this is not only an economic imbalance but above all a cultural one caused by a disparity in technological and information knowledge. This asymmetry, which is often referred to in the digital context as the *digital divide*, inevitably affects the correct formation of the will, even contractual, of the user. Indeed, at the time of the conclusion of an online contract, there is generally a considerable difference between the technological and information knowledge of the contracting parties ('information gap') that can cause erroneous expectations or unlawful reliance on the service provider so as to compromise the formation of the will.

This problem, well known to the European lawmaker, is amplified in the digital dimension where the big players of the web hold considerable economic power and have control over a huge amount of data and content which are used through techniques whose logics are often unknown or not transparent. This lack of transparency may unduly affect the users' transactional decisions and distort the normal competitive development of the market (Malgeri and Custers 2018).

It seems reasonable to assume that this context has probably led the European lawmaker to consider the web user as the weak subject in the relationships that take place online, thus allowing to draw a parallelism between the web user and the consumer. Not surprisingly, the recent Directive (EU) 2019/2161 for the better enforcement and modernization of Union consumer protection rules extends the scope of the Directive (EU) 2011/83 to contracts concluded between traders and consumers concerning digital services or contents.

This comprehensive perspective appears to be confirmed not only by the already examined legislative *corpus* proposed with the DSA but also by the rules enshrined in the GDPR. Indeed, the safeguards provided by the latter in favour of the data subject, namely, the natural person to whom the personal data refer, seem very similar to those previously established under Directive (EU) 2011/83 on consumer rights (Resta 2018). In both legislative frameworks, it is expected, for example, that information on the processing of personal data, as well as information relating to distance or off-premises contracts, must be provided through easily accessible methods and using simple, plain and intelligible language.

Likewise, consent to the processing of personal data is defined by the GDPR as "any freely given, specific, informed and unambiguous indication of the data subject's wishes by which he or she, by a statement or by a clear affirmative action, signifies agreement to the processing of personal data relating to him or her" (Article 4(1)(11) GDPR). Therefore, the GDPR seems to recall the characteristics of the contractual consent contained in the consumer protection rules, whereby it is possible to conclude a distance contract only after receiving all the information relating to the trader and the contract to be stipulated, and through the expression of prior express consent.

In the light of the abovementioned similarities, the nature of consent to the processing of personal data has been debated to understand whether it has the nature of authorization or negotiation (Resta and Zeno-Zencovich 2018; Twigg-Flesner 2016). Some scholars argue that the very definition of consent dictated by the GDPR would suggest that through the consent the data subject concludes a contractual agreement with the data controller. Other scholars, however, claim that consent to the processing of personal data would be a supplementary authorization, distinct and independent from any contractual agreement that contains it, thereby removing a limit to the power or authority that the law already grants to the data controllers to pursue their own interests.

Whereas the question of the nature of consent to the processing of personal data is still doubtful, what clearly emerges from the action of the European lawmaker is the will to protect, in a uniform manner and within different branches of the legal system, the weak subject or contracting party in a relationship that, from time to time, can be identified in the consumer, the data subject and, now, the web user.

5 Conclusions

Following the path already taken by the case law, the European lawmaker has aimed at strengthening the liability regime for digital providers. However, the goal pursued through the new set of rules established with the draft Regulation is not limited to make the digital dimension a safer and more transparent environment for web users. Indeed, it would be unwise not to take into account that the problem to be faced is broader and concerns the issue of sovereignty.

Although there cannot exist and will never be an Internet or an electronic commerce lawmaker, who dictates unique and unambiguous rules on the Internet or electronic commerce, the European lawmaker is trying to stem the overwhelming power of the big players in the digital market, such as Google, Facebook and Alibaba. Over time, these actors have built their rules, their systems, their dispute resolution methods, their private courts, exercising real sovereignty, not simply autonomy within the instrument of contract. In other words, in recent years the digital world is proving to be at the service of a new private sovereignty.

Against this state of affairs, the European lawmaker seeks to restore a hierarchical order in the digital world, establishing rules of public law nature to which the hitherto undisputed action of digital providers should be subordinated.

We are therefore navigating into an extremely complex regulatory framework, where rules having distinctive nature and corresponding to different kinds of sovereignty—from the public to the more private one—are intertwined. This may represent a starting point for reading the world where we live, a liquid society, as it has been authoritatively said by many (Bauman 2002), in which we certainly cannot ask for a solid law, but instead, we must deal with a law that is structured on multiple levels.

References

Bauman Z (2002) Modernità liquida. Laterza, Roma

Colangelo G, Maggiolino M (2019) From fragile to smart consumers: shifting paradigm for the digital era. Comput Law Secur Rev 35(2):173–181

De Franceschi A (2016) European contract law and the digital single market: current issues and new perspectives. In: De Franceschi A (ed) European contract law and the digital single market. The implications of the digital revolution, 1st edn. Intersentia, Cambridge, p 8–18

Delfini F, Finocchiaro G (eds) (2017) Identificazione elettronica e servizi fiduciari per le transazioni elettroniche nel mercato interno. Commento al regolamento UE 910/2014. Giappichelli, Torino

Finocchiaro G (ed) (2019) La protezione dei dati personali in Italia. Regolamento UE n. 2016/679 e d.lgs. 10 agosto 2018, n. 101. Zanichelli, Bologna

Hoffman DA (2016) From promise to form: how contracting online changes consumers. NYUL Rev 91:1595–1650

Malgeri G, Custers B (2018) Pricing privacy: the right to know the value of your personal data. Comput Law Secur Rev 34(2):289–303

Resta G (2018) Digital platforms and the law: contested issues. MediaLaws–Rivista di diritto dei media 1:231–242

Resta G, Zeno-Zencovich V (2018) Volontà e consenso nella fruizione dei servizi in rete (Will and Consent in the Provision of Services on the Internet). Riv Trimest Dirit Proced Civ 2:411–440

Twigg-Flesner C (2016) Disruptive technology–disrupted law? How the digital revolution affects (contract) law. In: De Franceschi A (ed) European contract law and the digital single market. The implications of the digital revolution, 1st edn. Intersentia, Cambridge, pp. 21–48

Authorities and Private Companies in Regulating Software Technologies

Vincenzo Ricciuto

1 From the Public Control of the Data to the Regulation of Its Flow

With the evolution of information technology and the spread of telematics, the relationship between public authorities and private companies in the context of the processing of personal data has been significantly changed. Consequently, the legal framework for understanding and regulating the phenomenon of personal data has also changed, though gradually. It still makes significant the question regarding the roles of authorities and private companies in regulating software technologies, considering that they are unstable over time. They change according to changes in the cultural sensitivities and objectives that a given community considers fundamental at a particular historical moment.

In fact, in a theoretical schematisation, we could identify a three-step structure of the relationship between public authorities and private entities in the regulation of data processing.

In a first step, the relationship is mainly seen as a relationship of conflict and opposition. The use of electronic personal data processing systems, which were initially expensive and complex and therefore widely used, especially by public authorities, poses the risk of penetrating control and recording of citizens by public bodies compared to the use of traditional paper records. For this reason, since the 1960s, national laws on data processing are aimed to provide guarantees for freedoms and personal rights in relation to databases and electronic archives held by public authorities.

In a second step, after the spread of electronic processors and information technology in common relations, there is a quantitative and qualitative growth in the

V. Ricciuto (✉)
Tor Vergata University of Rome, Rome, Italy
e-mail: vincenzo.ricciuto@uniroma2.it

R. Senigaglia et al. (eds.), *Privacy and Data Protection in Software Services*,
Services and Business Process Reengineering,
https://doi.org/10.1007/978-981-16-3049-1_2

dissemination of personal data and their use. At this stage, the issue regarding the protection of the rights of natural persons to whom the personal information refers concerns mainly the relations between private subjects, and, within these relations, impose the same requirements for guaranteeing and defending the right of the individual.

The role of public authorities is coherently redesigned. They are called upon to supervise the correct compliance with the guarantees and limits within which the law, on the one hand, allows an individual to process the personal data of other persons and, on the other hand, protects the person from the risks of such data processing. In the Italian regulatory framework, this function is particularly emphasised in the definition of the structure and tasks of the Italian Data Protection Authority provided by Law No. 675 of 1996 (which has implemented the first EU Directive 1995/46/EC on Data Protection). This independent administrative authority is vested with specific powers in order to protect the interested parties: such subjects may take an administrative action before the Authority, in the event of a violation of their rights in the context of the processing of their personal data. This represents an alternative to a judicial action (such a system has been defined as so-called 'double-track' protection).

Finally, there is a third and more complex stage where the processing of data is no longer considered as an activity of dangerous control of individuals that shall be limited and watched with suspicion and from which individuals shall be protected. The processing of personal data is also seen as a socially and economically profitable activity, a real driving force of a new economy: the so-called 'data-driven economy', which is particularly linked to the development of the digital environment and the Internet. The flow of personal data creates new forms of business, facilitates the exchange and supply of goods and services, and allows the functioning of the digital system: in this context, the processing of personal data and its flow represent values that the legal system seeks to protect, promote, and pursue.

Thus, the issue concerning the processing of personal data is legally structured and articulated (also) in an economic activity perspective by presenting the processing (also) as an economic phenomenon. This means that the subject of the processing of personal data, like all constitutive phenomena of economic activity, can be considered and regulated in accordance with the regulation of business and market activities, in addition to the classic defensive and restrictive approach. From this point of view, data processing opens up not only to the contractual perspective (because every company legally acts through contracts) but also to the idea that the flow of personal data can be organised according to the traditional rules of a market.

In such a context, the role of the sectoral competent authorities would coherently be enriched—in addition to the more traditional tasks of protection of interested parties—with tasks concerning market regulation and promotion of free movement of data.

On closer examination, these latter elements represent the reasons why EU Institutions have decided to regulate the processing of personal data since the first Directive of 1995. Nevertheless, the European Union has always moved towards contexts regarding the definition of a legal framework in relation to the free movement of personal data that is coherent with its objectives of free movement of goods, persons, services, and capital. These objectives are even more evident in the General Data Protection Regulation (EU) 2016/679 (GDPR): the choice of a European Regulation, which is directly and uniformly applicable in all Member States, is clearly due to the need to prevent different national implementations of the previous Directive that could constitute obstacles to the free movement of personal data (Recital 9 GDPR).

Nonetheless, the particularity of the issue and its unavoidable connection with the right of the individual preclude an untroubled declaration of an idea that there is a market of personal data which, in addition to the specific rules governing the personal guarantee, must also be subject to the rules governing business activities and related acts and contracts.

In Italy, for instance, the Regulation on the processing of personal data has been affected by an interpretation that has anchored it strongly to the issue of the protection of the person and secrecy.

It means that European regulatory intervention has been reductively considered as only a stage in the history of the recognition of the absolute rights, rather than as the beginning of the construction of the discipline of a particular market and the regulation of conditions and limits for the circulation of new forms of wealth. In this context, there has been a discussion about a "leading role of the interested party" within the phenomenon of personal data processing. It has overshadowed the negotiation profiles associated with the issues of data processing.

In this perspective, talking about a personal data market would have appeared not only eccentric but even dangerous in a hermeneutic and cultural context aimed to enhance aspects of an absolute nature that characterises fundamental rights over the aspects of a relative nature. Personal data is related to an individual, and it has long been argued that what relates to the person cannot constitute the subject of a market or a contract. Such a circumstance, however, means denying or misjudging the very principle of the free movement of data.

2 Personal Data Flow and Qualification of the Underlying Relationship

The resizing of the scope of the principle of the free movement of personal data at the national level, i.e., in the case of the implementation of Directive 95/46, resulted in the omission of any reference to the principle provided by Article 1 regarding the objectives of Law No. 675 of 1996.

Indeed, the only objectives mentioned in Article 1 of the Italian Act, which has implemented the aforesaid Directive on the processing of personal data, were those aimed at ensuring that "personal data are to be processed by respecting the rights, fundamental freedoms and dignity of natural persons, in particular with regard to privacy and personal identity".

The principle of the free movement of personal data, which is at the heart of the general Regulation, no longer allows us to assign any value to that reductive choice of the Italian legislator in the 1990s. On the contrary, efforts today must be made in the direction of a change of perspective, which shall be balanced with the fundamental need of the person, that would also raise the economic profiles of the data economy.

In fact, Article 1 of the GDPR states that the Regulation lays down rules on the free movement of data. Moreover, the data free flow within the Union shall neither be restricted nor forbidden for reasons connected with the protection of natural persons regarding the processing of personal data. Recital 6 of the GDPR highlights the way in which technology has transformed the economy and social relationships and thus also national legal systems, in order to increase the associated benefits. This should further facilitate the free movement of personal data within the Union and their transfer to third countries.

This requires emphasising, from a legal point of view, the activities through which the data flow takes place and, consequently, the subjective relationships underlying this phenomenon. This is a relationship between private subjects who carry out the movement of data which consists of the transmission and reception of information relating to natural persons. Following the right to data protection granted to the data subject, there also is the controller's right to process the personal data: in this dialectic, in the relationship between these two subjective rights, the public authority must be impartial and independent in relation to the parties and their interests.

Therefore, in a more traditional perspective, the role of the public authority is not only to guarantee or defend the involved individual but also to promote any activity related to the processing and regulation of the underlying interpersonal relations, especially in this recent phase.

Thus, the reinforcement of the principle of the free movement of personal data, highly emphasised by the GDPR, enriches what has been considered, for too long, as the only (defensive) *ratio* of the data protection regulation. Indeed, this also implies the need to regulate a new wealth of society based on information. The emphasis shifts from the protection of integrity which cannot be sold to the regulation of wealth capable of moving and feeding the economy.

It should also be noted that the national and European Authorities not only govern the activities of personal data processing and regulate their flow in accordance with the principles provided by the GDPR, but they are also constantly evolving their physiognomy in order to resemble other independent market regulators.

Thus, the Data Protection Authority not only established the legal boundaries of data processing and punishes possible damage suffered by individuals as a result of the processing of personal data but also guides processing activities by informing controllers or processors as to the changes which are necessary in order to comply with the provisions in force. An important role played by the Authority is to promote and adopt codes of conduct and professional ethics for specific sectors and to report to the Government the need for introducing legislative measures required by developments in the data protection field.

The GDPR has significantly expanded the powers of public authorities. They include typical measures of the administrative regulation of markets, such as the power to impose certain conditions of contracts concluded in regulated markets. In this manner, the authorities are able to control and guide the circulation phenomena of particular goods and services. In view of these contractual powers of public authorities, it should be stressed that the transfer of personal data to a third country or an international organisation may be carried out exclusively if it is in compliance with appropriate guarantees. In accordance with Recital 108 of the GDPR, the controller or processor shall compensate for the lack of data protection in a third country by providing appropriate safeguards to the data subject; these safeguards may consist of the use of binding corporate rules, standard data protection clauses adopted by the EU Commission, standard data protection clauses adopted by a supervisory authority, or contractual clauses approved by such an authority.

As mentioned above, the idea of the flow of personal data and the awareness of the spread of the data economy and their economic dimension lead to the conclusion that the instrument through which such flow takes place consists of a contract.

It does not lead to denying the protection of individuals in order to embrace the inescapable idea of its commodification. Rather, it could mean that in the case of processing personal data, the protection of a contracting party constitutes an additional element to the instruments of the protection of individuals. Moreover, in this situation, the involved individual maintains a series of rights that can be framed in a general supervisory power that is an expression of 'the right to data protection', which falls within more general protection of the personality. However, even in this case, consent—its manifestation—does not cease to be one of the essential elements of the data processing contract, as long as the interested party can dispose of their data in terms of enrichment and asset transfer, in accordance with the contractual schemes which provide a wealth circulation. This paves the way, for example, for the application of consumer protection even in the situation where an individual may dispose of their data in favour of a data collector.

3 The Contract and the Definition of the Models

The main theoretical and cultural difficulty is to admit that personal data could be the subject of a contract and thus could also be shared by an individual in exchange for a good or service.

Consequently, the problem of the specific regulation on economic transactions concerning personal data constitutes a relevant but less sensitive aspect.

However, it should be noted that the GDPR does not contain any discipline of the contract regarding personal data. The perspective of the GDPR is to identify the general rules for the processing of personal data, whatever the sector of reference may be; to identify the conditions underlying the data flow, such as the principle by which the flow occurs and whose origin comes from informed consent; and to define the rules and tasks of the market institutions of the data flow. Even if we assume that economic transactions may exist concerning the data, the GDPR still does not identify any specific kind of contract thereof and therefore does not provide any specific discipline for transactions in relation to an abstract scheme of services and rights on personal data.

Accordingly, such discipline for individual economic transactions and consequently for the contracts under which personal data flow occurs should be determined by EU law and, if necessary, national rules.

In this context, the provision of the GDPR must be integrated. The rules governing the digital service sector shall be integrated with the legal norms of Directive (EU) 2019/770 on certain aspects concerning contracts for the supply of digital content and digital services. Article 3(1) of this Directive considers, on the one hand, the contracts where a trader supplies or undertakes to supply digital content or digital service in exchange for a payment of a price as the subject matter of regulation and, on the other hand, the contracts where "the trader supplies or undertakes to supply digital content or digital service to the consumer, and the consumer provides or undertakes to provide personal data to the trader".

Once the subjective conditions are met, the specific protections for consumer contracts must be applied to all those cases where a natural person contractually provides personal data in order to obtain access to a specific good or service offered by the professional. On the other hand, in redefining the consumer policy, the European legislator considers the exchange of personal data for goods or services as a hypothesis substantially comparable to the traditional exchange of goods/services for money. Recital 31 of Directive (EU) 2019/2161 states that digital content and digital services are often supplied online under contracts where the consumer "provides data to the trader". It should be kept in mind that Directive 2011/83/EU applies to contracts under which the trader supplies digital services, and the consumer pays the price thereof. According to the aforementioned Directive of 2019, there is a need to extend the protections provided for the supply of digital services in exchange for money also to the cases of digital service contracts "under which the consumer provides personal data to the trader without paying a price. Given their similarities and interchangeability of paid digital services and digital services provided in exchange for personal data, they should be subject to the same rules".

With regard to the other contractual aspects and their discipline, it should be noted that the exact performance of the interested party concerning the personal data (and therefore the right acquired by the professional through the corresponding contract) is not established by law. There is no reference to the transfer of property rights over data (supposing that this is possible, nevertheless, this specific issue is still very much debated due to the inalienable nature of the personal aspects); there is no reference to the transfer of the right of use of these goods, such as a right to quiet enjoyment. There is even no limitation to the different uses of the personal data, i.e., productive use of the data subjects who 'sell' their data, productive use of the data controller (i.e., database that 'transfers' the data to third parties), or use as 'raw material' for the creation of further data, etc. Finally, no reference is made to an obligation to give (*dare*) or to do (*facere*) concerning the personal data. Therefore, the abstract scheme is open to different formulations.

The decision not to restrict (in an abstract prevision, the possible variables with which the phenomenon of data flow may occur) has a variety of reasons. Last but not least, it shall be highlighted that the new data-driven economy takes forms and delineations that are extremely flexible and variable and can very quickly suggest new forms and methods of flowing the good called 'personal data'. Forcing economic transactions into the abstract forms of predefined contractual schemes would therefore result, at this stage, contrary to the will of the EU regulatory framework, in terms of facilitating and feeding the new markets of personal data and the ever-rising forms of circulation of wealth.

4 The Data Market Affected by the Regulation

Ultimately, this modern interpretation of the phenomenon of personal data processing, interpreted in contractual terms as well, is due to the regulatory affirmation of the centrality of the principle of the free movement of personal data and to the fact that these data represent a new form of wealth that feeds the digital economy. Nevertheless, it must be stressed that, due to the persistence of theoretical and cultural difficulties, personal data may constitute an object of contract or trade which still precludes deducing from the premises (that all personal data have an economic value) all the consequences that could be deduced in terms of a protection increase and of efficient market regulation.

The greatest hostility is registered just within the national and European authorities called upon to regulate the sector of processing of personal data.

It will be sufficient to recall the role played by the European Data Protection Authority which amended the original draft of the Directive on certain aspects concerning contracts for the supply of digital content, by asking the European legislator not to qualify the transfer of personal data as counter-performance (see Article 3 of Dir. 2019/770). The 'nominalistic' measure, which did not change anything at the material level of the regulated phenomenon, was useful for the European Authority to save, from an ideological point of view, the idea that personal data cannot be considered as a mere commodity. Nevertheless, nowadays, data are treated and considered as goods with economic value. This is like saying that the data market exists and is regulated; however, it should not be called as such!

The same concerns of the Data Protection Authorities are not shared by other market regulators: the first one is the Italian Competition Authority (*Autorità Garante della Concorrenza e del Mercato,* further cited as the AGCM), which has a horizontal competence over the markets. In the Facebook case, which consisted in supervising Facebook's compliance activities concerning the regulations on consumer contracts in digital markets where the social network's services are offered to users in exchange for personal data (and not for money), the Italian Competition Authority has undoubtedly declared its competence and, consequently, has qualified these agreements as contracts for the exchange of digital services for personal data (AGCM, n. 27432/2018). The case in question concerned the information that Facebook provided to its users when they signed the contract: "Sign up! It's free and it will be forever". Such a slogan has been considered by the AGCM to be a misleading message and therefore in violation of the Unfair Commercial Practices Directive. Despite the fact that the contract did not require any monetary payment, it had to be considered an onerous contract in any case, as the service was provided in exchange for the user's personal data.

Consequently, the recognition of the existence of a contract concerning personal data, and thus their market, has allowed this particular data processing operation to be included within the scope of application of the consumer protection rules, paving the way for possible intervention by the competent authority.

In light of these brief considerations, it seems obvious that the issue regarding the relationship between public authorities and private companies in the regulation of software technologies faces a developmental boundary that depends on the appropriate conceptual structure of the idea of the personal data market. The most classic defence model of the individual with respect to the processing of personal data cannot exhaust the discussion of this specific issue. A proper balance must be found between the flow and promotion of the data market and the protection of the individual. In this context, it seems that the greatest effort shall come from a regulatory framework that is open to new phenomena and from the market regulators (the Competition Authority, the Authority for Communications Guarantees, etc.) who are called to regulate the processing of personal data due to their awareness of the data value in the markets.

References

Acquisti A, Taylor CR, Wagman L (2016) The economics of privacy. J Econ Lit 52(2):442–492

Autorità Garante della Concorrenza e del Mercato (2018), Provvedimento n. 27432/2018. https://www.agcm.it/dotcmsCustom/tc/2025/1/getDominoAttach?urlStr=192.168.14.10:8080/C12560D000291394/0/5428321DEA1A6FACC12584FC0050362B

European Data Protection Supervisor (2017) Opinion 4/2017 on the Proposal for a Directive on certain aspects concerning contracts for the supply of digital content. https://edps.europa.eu/sites/edp/files/publication/17-03-14_opinion_digital_content_en_0.pdf

European Data Protection Supervisor (2018), Opinion 8/2018 on the legislative package "A New Deal for Consumers". https://edps.europa.eu/sites/default/files/publication/18-10-05_opinion_consumer_law_en.pdf

Elvy SA (2017) Paying for privacy and the personal data economy. Columbia Law Rev 117(6):1369–1459

Gutwirth S, Leenes R, de Hert P, Poullet Y (2013) European data protection: coming of age. Springer, Dordrecht

Jerome W (2013) Buying and selling privacy: big data's different burdens and benefits. Stanford Law Review Online 66(47)

Langhanke C (2018) Daten als Leistung. Mohr Siebeck, Tübingen

Langhanke C, Schmidt-Kessel M (2015) Consumer data as consideration. EuCML 1:218

Lohsse KS, Schultze R, Staudenmayer D (2017) Trading data in the digital economy: legal concepts and tools. Oxford University Press, Oxford

Lynskey O (2015) The foundations of EU data protection law. Oxford University Press, Oxford

Organisation for Economic Cooperation and Development (2018) Quality considerations in Digital Zero-Price Markets, background note by the Secretariat. https://one.oecd.org/document/DAF/COMP(2018)14/en/pdf Accessed 20 Mar 2021

Prins C (2006) When personal data, behavior and virtual identities become a commodity: would a property rights approach matter? 3(4):270–303

Rhoen M (2015) Big data and consumer participation in privacy contracts: deciding who decides on privacy. Utrecht J Int Eur Law 31(80):51–71
Ricciuto V (2020) Il contratto ed i nuovi fenomeni patrimoniali: il caso della circolazione dei dati personali. Rivista di Diritto Civile 3:642–662
Ricciuto V (2018) La patrimonializzazione dei dati personali. Contratto e mercato nella ricostruzione del fenomeno. Il Diritto Dell'informazione e Dell'informatica 3–4:689–726
Robertson VHSE (2020) Excessive data collection: privacy considerations and abuse of dominance in the era of big data. Common Market Law Rev 57(1):161–190
Schwartz PM (2003) Property, privacy, and personal data. Harv. L. Rev. 117(7):2056–2128
Senigaglia R (2020) La dimensione patrimoniale del diritto alla protezione dei dati personali. Contratto e Impresa 2:760–783
Solinas C (2021) Circolazione dei dati personali, onerosità del contratto e pratiche commerciali scorrette. Giurisprudenza Italiana 2:320–335
Stucke ME, Grunes AP (2016) Big data and competition policy. Oxford University Press, Oxford
Synodinou TE, Jougleux P, Markou C, Prastitou T (2020) EU internet law in the digital era: regulation and enforcement. Springer, Cham

Liability and Accountability in the 'Digital' Relationships

Carmelita Camardi

1 The Complexity of the Liability Regulation Within the Digital Relationships in the EU Law and the Domestic Law

It is well known that the legal regulation of the so-called 'digital relationships' has moved towards different directions in the European landscape, not only for the rapid and multiple dynamics of Industry 4.0 but also due to the change of the regulatory strategies developed in the European and national debates towards those dynamics, in particular on the perspective that has to be taken and the objectives to be pursued.

The complexity of the system arises due to the growing instability of EU law, in addition to the instability brought by urgent solutions that occur from the disputes set out before the Court of Justice, as happened, for example, in the application of the GDPR to the transfer of personal data to non-EU countries (among others, the 'Schrems cases').

Another element of complexity derives from the interaction between European and national laws: a physiological interaction, considering that for the concrete application of the European principles and rules both directives and regulations require an adaptation within the entire national system, including its specific disciplines and its technical and legal culture. However, with respect to EU law, and, despite its primacy and consistent unifying force, there are always potentially controversial cases within domestic jurisdictions.

In the following brief considerations, the aforementioned problem will be considered with the liability of data controllers and data processors towards data subjects, as mainly set out by Article 82 of the GDPR. Furthermore, specific aspects of liability, which are regulated by Directive 770/2019/EU on certain aspects concerning

C. Camardi (✉)
Ca' Foscari University of Venice, Venice, Italy
e-mail: camardi@unive.it

R. Senigaglia et al. (eds.), *Privacy and Data Protection in Software Services*, Services and Business Process Reengineering,
https://doi.org/10.1007/978-981-16-3049-1_3

contracts for the supply of digital content and digital services, should be recalled also. Also, the new 'Acts' under discussion by the EU institutions will be contemplated, which aim to govern the activities of digital platforms (DSA and DMA on certain behaviours undertaken by platforms acting as digital 'gatekeepers' has been presented in December 2020).

It should be noted that the rules set out in the GDPR are very broad since the Regulation systematically covers the processing of personal data carried out by any subject. Therefore, it is intended for its general application for any operations in which the processing of personal data is considered, either expressed or implied, either for the commercial or the public relationship between the natural person and the subject who, while carrying out an activity of any kind, ends up with collecting personal data. This means that many other rules, such as those applicable to the (commercial) relationships between professionals and consumers, may be integrated and added to the data protection law under the GDPR by adapting them as a 'special' legal framework.

2 The Liability of the Data Controller and the Principle of Accountability. Article 82 of the GDPR

Article 82 states that "Any person who has suffered material or non-material damage as a result of an infringement of this Regulation shall have the right to receive compensation from the controller or processor for the damage suffered". After having shared the liability between the controller and the processor, par. 3 affirms that: "A controller or processor shall be exempted from liability under paragraph 2 if it proves that it is in no way responsible for the event giving rise to the damage". Finally, par. 6 declares that "Court proceedings for exercising the right to receive compensation shall be brought before the courts competent under the law of the Member State referred to in Article 79(2)". This could be seen as an express reference to the national law of each Member State.

Many other provisions, together with several recitals, enrich the current analysis with some important content. First of all, the definition of the concept of a personal data breach, as set out in Art. 4(1)(12) GDPR, outlines "a breach of security leading to the accidental or unlawful destruction, loss, alteration, unauthorised disclosure of, or access to, personal data transmitted, stored or otherwise processed". Furthermore, Recital 75 highlights the damages which may result from unlawful data processing, in particular: "discrimination, identity theft or fraud, financial loss, damage to the reputation, loss of confidentiality of personal data protected by professional secrecy, unauthorised reversal of pseudonymisation, or any other significant economic or social disadvantage". This Recital continues by mentioning that the damage that may be caused to the rights and freedoms of individuals where they are "prevented from exercising control over their personal data; where personal data are processed which reveal racial or ethnic origin, political opinions, religious or philosophical beliefs, trade-union membership, and the processing of genetic data, data concerning health or data concerning sex life or criminal convictions and offences or related security measures; where personal aspects are evaluated, (...) in order to create or use personal profiles; where personal data of vulnerable natural persons, in particular children, are processed (...)". Moreover, Recital 85 reiterates the harmfulness of a personal data breach, as the controller should notify the supervisory authority generally without undue delay.

In a nutshell, the previously mentioned provisions highlight the awareness of EU institutions over the (high) risks related to the processing of personal data, as well as the (severe) potential offence to the rights of individuals caused by digital technologies, in which some of the above-mentioned damages can be found, such as the reversal of pseudonymisation or the undue profiling.

Given this awareness, the general model of liability outlined by Art. 82 GDPR seems to be more than justified, especially where it links the obligation to the compensation of damages for any violation of the GDPR, whatever it may be, hence the whole regulation appears to be functional to the protection of information concerning an individual, whether such data have patrimonial contents or not.

Before addressing the problems linked to Art. 82 of the GDPR in the context of national legal systems and specifying the interpretation that has to be provided, it is necessary to add some more consideration.

The rules that give data processing the structure of an activity with a highly technical capacity are synthetically encapsulated in the so-called 'principle of accountability', in which the GDPR clearly states the importance of a prior impact assessment of the processing over data protection to deal with the chances and gravity of risk, especially when a certain type of processing "is likely to result in a high risk to the rights and freedoms of natural persons" (Sect. 3, Art. 35 et seq.). Otherwise, it requires the controller and the processor to implement appropriate technical and organisational measures to ensure the level of security proportionate to the risk, and, first and foremost, to ensure and be able to demonstrate that processing is carried out in accordance with the provisions of the GDPR (Art. 24).

More specifically, Art. 32 states that "the controller and the processor shall implement appropriate technical and organisational measures to ensure a level of security appropriate to the risk, including inter alia as appropriate": pseudonymisation, ability to ensure the confidentiality, integrity of processing systems, ability to restore the availability and access to personal data. While Article 25 provides that "both at the time of the determination of the means for processing and at the time of the processing itself, implement appropriate technical and organisational measures, such as pseudonymisation, which are designed to implement data-protection principles, such as data minimisation, in an effective manner and to integrate the necessary safeguards into the processing in order to meet the requirements of this Regulation and protect the rights of data subjects" (data protection by design and by default).

More broadly, the principle of accountability outlines the structure of the processing of personal data and the role of the data controller according to a model in which it takes the danger and risk of data processing into account, as well as considers the preventive technology and the 'security program' as an antidote against the dangers and damages that the same digital technologies might determine. From this point of view, the punctual regulation of the processing in all its steps, under the control of independent supervisory authorities, includes mandatory rules that represent, alongside the regulation of the free movement of personal data, a specific declination of the individual economic freedom in the digital market, as such offered to the freedom of organisation of digital players in the data-driven economy.

In this manner, it has been paved the way to investigate the main topic of this article, which is enclosed between liability and accountability in the perspective of EU law.

On a terminological level, 'liability' refers to a legal duty or obligation, i.e. the condition of who is legally responsible for something (either because they have to do it, or because they have to pay for it), and this is the term used in the Art. 82(2) ("Any controller involved in processing shall be liable for the damage caused by processing (…)").

Otherwise, 'accountability' refers to the condition of the person who is responsible for a decision or action and, if requested, they must be able to demonstrate their activity. The word is used in Art. 5(2) of the GDPR to define, in principle, the role of the data controller, who "shall be responsible for, and be able to demonstrate compliance with, paragraph 1 (accountability)"; it is also recalled in Recital 85 (about the notification of a personal data breach), and it conveys the idea of 'taking into account' something that has been done to fulfil an obligation in this way. For example, the principle of accountability justifies the recommendation to the data controller to acquire the written consent of the data subject, in order to demonstrate it, if requested, since the controller is 'accountable' for the acquisition of consent when it is the legal basis for processing.

Both terms 'liability' and 'accountability', therefore, evoke the situation of a subject who is obliged to behave in a way that is functional to the enforcement of the rights of data subjects; then, while each locution characterises a different juridical position within the relationship between the data controller and its interlocutors, those positions may potentially interfere with each other's.

More specifically, 'accountability' defines a proper organisational aspect on the ongoing data processing, since the data controller must be able to demonstrate the compliance, i.e. the proof of having adopted appropriate technical measures to guarantee the data security and the protection of the rights of data subjects concerning the obligations of the Regulation (Art. 5 GDPR). However, proof of compliance may be required irrespective of the existence of a harmful event or a data breach, as well as for obtaining consultation or prior authorisation for certain types of processing (see Articles 35 and 36 GDPR). In those cases, the expression 'responsibility' also is useful to indicate the duties and obligations that physiologically arise from the data processing as a dangerous activity that requires specific 'care' for being performed. Accordingly, the locution 'responsibility' describes what—in other words—is defined as a risk-based approach, which stays for the preventive approach that the EU law has adopted. Nevertheless, Art. 24 of the GDPR is significantly dedicated to the 'Responsibility of the controller', and it affirms in par. 1 that "the controller shall implement appropriate technical and organisational measures to ensure and to be able to demonstrate that processing is performed in accordance with this Regulation".

On the other hand, 'liability' is referred to what—in the Italian legal tradition—is the concept of 'civil liability', as an institution that regulates the consequences of a harmful event for the subject who has to pay for it (duty to compensate for damage), by following a legal criterion of imputation.

In this situation, 'accountability' and 'liability' may interfere, and differently shape the 'responsibility' burdening on the data controller, depending on the type of damage that has occurred to the rights of data subjects.

This is what is possible to deduct from Art. 82(3) of the GDPR (which contains the word 'liability') according to which "[a] controller or processor shall be exempt from liability under paragraph 2 if it proves that it is not in any way responsible for the event giving rise to the damage". Thus, for not harming the effectiveness of Articles 35 and 36 of the GDPR and many other rules which substantiate the principle of accountability, it cannot be left behind the concrete adoption of the technical and organisational measures in which the accountability is expressed; the same principle also comes into force when the data controller is asked to compensate for the damage—for example, a data security breach—when the rigorous attendance to those measures would have avoided.

In essence, if the damage claimed by the data subject concerns their right to data security, the nature of the evidence under Art. 82 would already be predetermined in certain aspects, or rather, since it requires to prove that the harmful event is not linkable to the controller "in any way", it also implies that the latter can exhibit any certification or documentation adopted early concerning the selection of security measure—based on a prior assessment on a risk-based approach—suitable to prevent the damage suffered by the data subject. Therefore, the data controller shall be liable for the damages resulting from the failure of adopting measures that are technically possible and proportionate to the outcome of the data protection impact assessment (see Article 35 GDPR). Otherwise, the controller shall not be liable for the damages that have been excluded from the assessment, if the test has been properly carried out, because they were not foreseeable, highly improbable or not remediable by the state of the art. In essence, the data controllers might be exempted from their responsibility/liability by proving that they have carried out a complete risk assessment, as well as having adopted the related architecture of the processing in accordance with the purpose and the means of it in order to avoid the causation of such damages (full compliance with accountability principle) and, however, the damage may occur as a result of unforeseeable circumstances that could not be estimated beforehand.

3 The Nature of Responsibility Under Art. 82 GDPR

How accountability and liability characterise the processing of personal data is deemed beneficial to verify the interconnection between the EU law and the national law, in particular the Italian law.

The main issue comes from the above-mentioned general nature of the liability outlined by Art. 82 GDPR, while in the Italian legal tradition, there are two different types of liability: the civil liability pursuant to Art. 2043 et seq. of the Italian Civil Code and the liability for breach of contract under Art. 1218 et seq. Ital. Civ. C. The distinction between the two species, as it is well known for Italian legal scholars—but also in common law systems—includes not only the elements of liability and the duty to compensate for damage but also the rules which apply to each figure for some important aspects, such as the following: the kinds of compensable damages (limited by Art. 1225 to that of foreseeable damages at the time the obligation arises); the nature of the compensable damage (only pecuniary damage, while it is still disputed the compensation for non-material damage when the creditor had failed to comply with its obligation); the statute of limitations; the proof of loss or the exonerating proof.

Art. 82 of the GDPR could not provide specific indications with respect to most of those elements, while this regulation refers to the provisions of national laws. Here the problem of the 'qualification' of liability is not only theoretical but also empirical. Italian scholars have frequently disputed the 'nature' of liability—contractual vs. extra-contractual—ascribed to the data controller for the damages caused by the violation of the rules on the processing of personal data.

In a nutshell, if Art. 82 of the GDPR is interpreted as a rule aimed at regulating the extra-contractual liability of the controller, it would be confirmed by considering that Art. 82 succeeded the former Art. 15 of the Italian Privacy Code, by which whoever causes damage to another as a consequence of the processing of personal data, shall be liable to pay damages pursuant to Art. 2050 Ital. Civil C. Moreover, both provisions refer to the non-patrimonial loss (defined also as non-material damage), which evokes the context of non-contractual liability and the provision of Article 2059 Ital. Civil C. Otherwise, in the same direction, the exonerating proof—as the data controller is not in any way for the event giving rise to the damage (Art. 82 GDPR)—seems to regulate, insofar as it is mentioned an "event", the allocation of risk relating to the performance of an activity. Thus, it appears similar—although not coincident—to Art. 1218 Ital. Civil C. for which the debtor is exonerated from the liability for non-performance of the obligation, while the non-performance of the duty was caused by the impossibility resulting from a cause not attributable to them.

From the opposite point of view of those who interpret the Art. 82 of the GDPR as regulating the contractual liability of the controller, it should be considered an aspect of both textual and systematic value. The explicit reference to the damage "as a result of an infringement of this Regulation" is a factual element of the liability, insofar that the damaging effects to the rights of data subjects can be ascribed to the data controller as a result of a violation—no matter if blameless—of the rules set out for the processing of personal data. The only exception is when the controller is not responsible in any way for the event giving rise to the damage.

The meaning of Art. 82 of the GDPR lastly recalled may enhance systematically the whole aspects of the principle of accountability, which is embodied in the appropriate technical and organisational measures requested to ensure the highest security in the processing of personal data and, in particular, concerning the peculiarities of the undergoing processing (by taking into account the nature, scope, context and purposes of data processing). Furthermore, the fulfilment of those obligations does not apply only in the context of the determination of the processing as an activity supervised by public authorities but also for the obligations and duties that can be claimed by data subjects against the data controller who processes their data. In essence, the duty of data security in the storage and the processing of data, as well as other legal obligations (e.g. the methods of acquiring consent when consent is the legal basis for processing and many other duties pursuant to the GDPR), despite their legal origin, should establish an obligatory relationship between the controller and the data subject, as a result of a general phenomenon of legal determination of contents on the latter relation. Hence, the factual aspect of processing would be capable of generating duty to act (to do and not to do something, etc.) for the data controller to protect the rights of data subjects. Accordingly, the qualification of liability pursuant to Art. 82 of the GDPR as a liability for breach of obligations, which is based on the model of Art. 1218 Ital. Civ. C., regardless of the existence of a contractual relationship between the data controller and the data subject.

This argument, related to the interpretation of Art. 82, requires some further considerations.

First of all, the regulatory philosophy chosen by EU institutions and adopted in the GDPR does not reflect pre-established and dogmatic legal categories. This is not only due to the application of the Regulation to multiple institutional subjects within their legal systems but also by the nature of the regulated activity, as well as the strong public insight of the GDPR, notwithstanding the explicit statement of the free movement of personal data as an indispensable pillar of the EU market.

In essence, the processing of personal data may be considered as a transversal activity of the Digital Single Market, as well as of the digital society, by considering that the spread of digital technologies is very broad, moving from the domestic to the economic environment, from private to public relations; nevertheless, such technologies require a basic and unitary discipline that can be operational and enforced both by public and private actors.

Moreover, as a normative aspect that testifies the level of legal civilisation existing in the EU, data processing is defined by the GDPR as an activity that always interferes with the fundamental rights of the individuals, which should be protected in any case, albeit balancing them with the free movement of data. This is confirmed—as it was mentioned at the beginning of this essay—by the discipline of the supply of digital content and digital services, which establishes the primacy of the GDPR on data protection over the contracts that perform operations over personal data of the consumer, as ruled by the Directive (EU) 770/2019, even before the controller has obtained valid consent.

In this respect, data protection and duties of data security under the GDPR automatically become part of the contract between the supplier and the consumer, and the latter may claim compliance with the data protection law albeit those obligations were not mentioned in the binding agreement or the trader has excluded them.

If the processing of personal data, in any case, must be carried out in respect of the rules laid down by the principle of accountability in all its implications, and if those rules aim to protect the fundamental rights of the individual whenever it is the legal basis for processing (consent or another lawful basis), it would have been an interpretative error to consider Art. 82 of the GDPR as a rule that covers only the contractual liability of the data controller or, for the same reason, only the extra-contractual liability. In other words, this is nonsense.

Data protection compliance and the subsequent liability for any damage caused by an infringement of the GDPR should always be ascribed to the subjects mentioned in Art. 82 of the GDPR, except for the cases in which the controller and the processor prove that they are not in any way responsible for the event giving rise to the damage. This discipline fulfils its rationale as it is translated by the data controller into the specific adoption of the technical and organisational measures, as well as for security tools in order to face the risk of a data breach, and the other duties aimed at the same purpose (the duty to provide information to the data subjects, also if requested by the data subject who exercise its rights of access, to rectification, etc.).

What is different are the modalities through which the data subject could claim its rights and eventually the compensation for damages in the two different situations.

(a) If the processing is carried out within a contractual relationship, the data subject/consumer/party to the contract is enabled to exercise the remedies set out by the GDPR whenever the conclusion and/or the performance of the contract has required the processing of personal data of the consumer/data subject to the professional. In this respect, the duties settled by the GDPR for the data controller will be enforceable as well as the contractual clauses, which are integrated with the data protection law. In the case of damage, the consumer/data subject will be able to claim compensation according to the model of contractual liability (Art. 1218 et seq. Ital. Civ. C.), as well as to claim compensation for non-pecuniary damage.

(b) If the processing is carried out without a contractual relationship between the data controller and the data subject, the controller, however, shall be liable under the GDPR for the damages caused to the rights of the data subject, as such damages derive from an infringement of the Regulation, regardless the possibility for the data subject to pretend the duties of data security or disclosure, which have to be provided in any case because they pertain to data processing, even before being considered as necessary content of a contractual relationship. Thus, if personal data are processed while the data subject itself is carrying out an activity, for example, recorded by a video surveillance system installed in a shopping centre, the data subject is enabled to exercise their right to compensation for damages (material and non-material damage) against the controller who processes the data captured by the camera, if there are any harmful consequences as a result of the infringement of the GDPR, according to the model settled down by Art. 2043 and 2050 Ital. Civ. C. The owner will be exempt from liability only by demonstrating the exonerating proof of Art. 82(3) of the GDPR.

Legal scholars have already perceived the wider scope of application provided by Art. 82 of the GDPR. It has been suggested that the qualification of Art. 82 of the

GDPR is as an 'ancipital' model, which stays in the middle of liability for breach of obligations and extra-contractual liability. This argument is not convincing, at least from the point of view of the legal categories, as a synthetic content of the discipline. On the contrary, it may offer an acceptable solution from the viewpoint of regulatory efficiency, i.e. the application of the data protection law by letting the individual exercise their right to data protection and control of their data in any situation in which, voluntarily or accidentally, a data processing is performed.

At the Italian domestic level, the 'ancipital' responsibility laid down by the Art. 82 comes from a factual element ruled by the law as producing duties pursuant to Art. 1173 Ital. Civ. C., which has identified the sources of obligations (contract, tort and "any other fact suitable to producing them in conformity with the legal system"). Therefore, the rigid distinction between the contractual and extra-contractual liability does not suffice, at least for identifying the discipline applicable to duties and the remedies for their non-performance. This has emerged in the literature before the movement of personal data (in a digital context); the data flow, however, has highly stressed the traditional legal categories of civil law. Among others, the issues generated by the pre-contractual liability or 'social contact' should permit us to understand the complexity of the institution of civil liability to face the promiscuous circumstances that may give rise to harmful events.

A similar complexity—and maybe even greater—may characterise the current movement of personal data, which is subject to conflicting principles to be balanced and business models that lie in the middle of the public and private sphere, as normally happens in economic sectors subject to a regulatory philosophy.

In this respect, the dialectic between accountability and liability, which is typical of the choices made by EU institutions, as well as a 'polyfunctional' model of liability (*ex ante* and *ex post*), shows the complexity of the plural systems of the present age.

References

Alpa G (2019) La proprietà dei dati personali. In: Galgano NZ (ed) Persona e mercato dei dati. Riflessione su GDPR. Cedam, Padova

Amore G (2020) Fairness, Transparency e Accountability nella protezione dei dati personali. Studium Iuris 4:414–429

Barbierato D (2019a) Trattamento dei dati personali e "nuova" responsabilità civile. Responsabilità Civile e Previdenza 6:2151–2159

Bilotta F (2019b) La responsabilità civile nel trattamento dei dati personali. In: Panetta R (ed) Circolazione e protezione dei dati personali, tra liberà e regole del mercato. Commentario al Regolamento UE n. 2016/679 (GDPR) e al novellato d.lgs. n. 196/2003 (Codice Privacy). Giuffrè Francis Lefebvre, Milano

Bravo F (2018) Sul bilanciamento proporzionale dei diritti e delle libertà "fondamentali", tra mercato e persona: nuovi assetti nell'ordinamento europeo? Contratto e Impresa 24(1):190–216

Bravo F (2019) L'«architettura» del trattamento e la sicurezza dei dati e dei sistemi In: Cuffaro V, D'Orazio R, Ricciuto V (eds), I dati personali nel diritto europeo. Giappichelli, Torino

Bravo F (2019) Riflessioni critiche sulla natura della responsabilità da trattamento illecito di dati personali. In: Galgano NZ (ed) Persona e mercato dei dati. Riflessione su GDPR. Cedam, Padova

Camardi C (2019) Prime osservazioni sulla Direttiva (UE) 2019/770 sui contratti per la fornitura di contenuti e servizi digitali. Operazioni di consumo e circolazione di dati personali. Giust Civ 3:499–523

Cuffaro V, D'Orazio R, Ricciuto V (eds) (2019) I dati personali nel diritto europeo. Giappichelli, Torino

Finocchiaro G (2019) Il Principio Di Accountability. Giurisprudenza Italiana 12:2778–2782

Mantelero A (2019) Gli autori del trattamento dati: titolare e responsabile. Giurisprudenza Italiana 12:2799–2805

Ratti M La responsabilità da illecito trattamento dei dati personali nel nuovo Regolamento. In: Finocchiaro G (ed) Il nuovo regolamento europeo sulla privacy e sulla protezione dei dati personali. Zanichelli, Bologna

Riccio GM (2018) Diritto al risarcimento e responsabilità. In: Riccio GM, Scorza G Belisario E (eds) GDPR e normativa privacy. Commentario. Giuffrè Francis Lefebvre, Milano

Ricciuto V (2018) La patrimonializzazione dei dati personali. Contratto e mercato nella ricostruzione del fenomeno. Il Diritto Dell'informazione e Dell'informatica 4:689–726

Thobani S (2019) Il danno non patrimoniale da trattamento di dati tra danno presunto e danno evento. Giurisprudenza Italiana 1:43–46

Torino R (2019) La valutazione d'impatto (Data Protection Impact Assessment). In: Cuffaro V, D'Orazio R, Ricciuto V (eds), I dati personali nel diritto europeo. Giappichelli, Torino

Tosi E (2020) La responsabilità civile per trattamento illecito dei dati personali alla luce del General Data Protection Regulation (GDPR). Studium iuris 7/8:840–845 and 9:1032–1038

Social Media, Mobile Apps and Children Protection

Roberto Senigaglia

1 The Role of the 'Environment' for Child Development

The impact of digital technology in the personality development of a person under the age of 18 assumes particular aspects since it is generated by algorithm techniques, which operate with data and depicts social relations as continuous and insatiable 'input-output' mechanisms, away from the attention and empathy of interpersonal relationships traditionally recognised to minors. Accordingly, it is necessary to rethink the remedies and categories traditionally used to protect children, and, more broadly, to change the methodological perspective in the analysis of familial and social relationships that have to do with the minor, as well as the related solutions.

Considering the issues of social networks usage among minors, as well as when they search for information in the web ecosystem, facing the pitfalls of the digital market, the lack of adequacy of legal rules and rigid solutions could be seen, which traditionally pertain to a formal and sectoral model of civil law systems. On the contrary, it is necessary to put together both private and public law enforcement, which are capable of considering the individual unitedly, from the most intimate sphere to patrimonial aspects, as well as their civic and social role.

It is well known that special consideration has been recognised for children by the 'patchwork' of sources of European and international law, as well as by different forms of soft law: not so much for the fact of age, while the minor is particularly vulnerable in the first years of life—at least until the capacity for discernment—rather that child development is strictly linked to the evolutionary path, which makes up individual personality and self-awareness towards adulthood. The 'role'

R. Senigaglia (✉)
Ca' Foscari University of Venice, Venice, Italy
e-mail: robseni@unive.it

R. Senigaglia et al. (eds.), *Privacy and Data Protection in Software Services*,
Services and Business Process Reengineering,
https://doi.org/10.1007/978-981-16-3049-1_4

of law to the latter step is the achievement of a certain level of maturity and awareness/consciousness, which is required for embracing the full (free) capacity to act and exercising the right of individuals.

The importance of child development for shaping personal identity underlines the special attention accorded by legal rules to the environment in which the minor grows up, by arranging, guaranteeing, and monitoring the 'warranty of suitability' to provide the person with the fundamental values of the legal system. In essence, also the 'location' is functional to the affirmation of human dignity, where the rights culture and the fundamental aspects of legal humanism cannot be denied. Against those fundamental issues, some situations could expose the child to the denial of relationality or familial relations, solidarity, emotional sphere, by abandoning him to solitude, the muteness of affection, isolation, and hatred.

In the same direction, there are the principles of international origin that govern juvenile law: on the one hand, the primary consideration of the best interests of the child (Article 3, United Nations Convention on the Rights of the Child, furthermore cited as the CRC), which sets out the criteria to be taken while interpreting and choosing solutions over relations of subjects under the age of 18; on the other hand, the right to be heard, as well as the right to express their interest, which has been recognised to the child who is capable for discernment, i.e., who has reached the maturity that allows them to distinguish the choices that are for or against their interest. With particular regard to the latter, Article 12 of the CRC, after having recognised the child capable for discernment of the right to express those views freely in all matters affecting themself, affirms that the views of the child being given due weight in accordance with the age and maturity of the child.

Those principles are reaffirmed by Article 24 of the Charter of Fundamental Rights of the European Union (the only article dedicated to the rights of the child), which states the right to protection and care as is necessary for their well-being, thus directing the educational action (care) of those who have a familial and institutional relationship with them, as well as all actions and instruments for the protection of their person, per the best interest of the child.

2 The Contribution of Education

In the process of maturing from a child to an adult, education plays a decisive role because it moulds the values of the individual personality. Based on national constitutional charts, as well as the CRC, the right and duty to educate the child is, naturally, parental responsibility; insofar that it is a legal situation connected to the procreation (and not to the filiation), and it affects the procreative responsibility, for which the parent has the duty of care the child to carry out under the deontic triad of education, instruction, and maintenance.

The general principle that traditionally governs the exercise of parental responsibility is the neutrality from any form of interference by the outside, with the only limit to comply with the best interests of the child; thus, only a conflicting interest may authorise the general application of this rule and the subsequent demand for forms of external intervention aimed at protecting the child.

In essence, under the CRC, each contracting Party shall guarantee, with the necessary positive and negative actions, the principle that both parents have common responsibilities for child development; otherwise, States shall refrain from interfering with this responsibility as long as it is based on the best interests of the child (Article 18). Moreover, when the education seems to be characterised by other purposes, the State intervention is not only justified but essential to permit, in compliance with the international obligations, the interest of the child to be fully prepared "to live an individual life in society" and brought up "in the spirit of the ideals proclaimed in the Charter of the United Nations, and in particular in the spirit of peace, dignity, tolerance, freedom, equality and solidarity" (CRC Preamble).

In a nutshell, due to the pattern that emerges from the abovementioned regulation concerning the relationship between parents and children, both education and care are addressed to the parents and ruled by freedom of choice, which is limited by the best interests of the child. Several actors may intervene to support the parents (school, sports club, religious associations, etc.) if they consent and in any case under their control, at least until when the child is capable of discernment.

This scheme presupposes—as it was before the digital era—the exclusion of forms of interference and support at the educational level except for parental control.

In today's reality, technologies and Artificial Intelligence (AI) have become both aspects of the environment where family life occurs—having changed individual relationships, especially in terms of mutual dedications—and elements of the people who live in such places. Therefore the interpersonal relationship may constrain, even in an invasive way, the connection between the individual and the digital tools; even though such technologies—despite the former—do not work as interlocutors but as receivers and sources of information and virtual relations (which are no relations at all).

The unprecedented factor, therefore, cannot be seen only in the usage of technology in the context of a familial relationship as a simple support to the parental responsibility; rather it consists in the assumption of an immediate and direct role in the personal development, by providing an environment '*prêt-à-porter*' to the child, which allows them to seek refuge in it, by using means that easily are not under the control of the other family members (especially the parents) and therefore open up to uncontrollable effects.

By looking at this new reality, scholars recently have spoken about the 'Digital Family'. While they underline the threats of the Web, characterised by information, social and economic operations, for the family and, dynamically speaking, for familial relationships; in particular, on the one hand, technologies have established communicative/educational actions which are parallel and at the same time hidden to the parental control; on the other hand, the minor is left to relationships even more formal and sterile, which have nothing to do with the real and social life.

3 Hard Law and Soft Law

The pervasive force of the personal and relational sphere of minors, which characterises digital technology (without ethics), requires the legal system to provide adequate answers both in terms of protected interests (the emergence of new rights) and applicable remedies, by recurring not only to the dimension of subjective rights and sanctions but also to the regulatory private law together with its preventive functions.

The reference is intended to forms of self-regulation and co-regulation, mainly to be implemented through codes of conduct.

In the processing of personal data, where the traditional means do not suffice to guarantee effective data protection, the use of the above-mentioned self-regulation tools is encouraged especially for associations and other bodies representing categories of data controllers or data processors, in order to facilitate the effective application of the Regulation (EU) 2016/679 (also known as GDPR) "taking account of the specific characteristics of the processing carried out in certain sectors". Such rules, as a result of private autonomy, are enforced, even at high levels of generality, by the control and approval of the supervisory authorities (Articles 40 and 41 GDPR). In this same problematic field, expressions of private regulatory law, functional to the effectiveness of rights, are also the rules concerning technical and organisational measures, which shall be implemented by the data controller to fulfil data protection by design and privacy by default; those rules should take into account, among other things, also "the risks of varying likelihood and severity for rights and freedoms of natural persons posed by the processing" (Article 25 GDPR), by considering that high level of risks certainly could be reached when dealing with data relating to individuals under the age of 18.

The reference to soft law regulation is undoubtedly necessary to neutralise the risk of failure of parental educational attainment.

From the earliest years of life, children start to keep confidence, even at home, with smart devices, which entertain them with stories, images, games, etc. As soon as children develop social skills, the world of social networks attracts them—sometimes even 'consumes' them—becoming an indispensable place for their social life, as far as switching off the smartphone makes them feel disconnected from the world. Here the vulnerability of the person under the age of 18—the most ethically and culturally developed one—is exposed to higher risks, by considering how those 'social formations' are represented together with personal data, which allow the smart devices to build up an 'artificial' identity of the data subject outside the control of the children and their parents.

This background was unknown by the aforementioned sources of juvenile law in the last century. Also, the role of parents in education and child development and protection, especially in the first years of life, is no longer sufficient alone but shall be complemented by the actions of other subjects. In essence, the entrance to digital social life by the side of children, as well as the access to information that they can take from the social media and the whole Web outside of the parental control demands new educational roles: not only the subjects who remain in a relationship with their parents (e.g., schools and social services) but also private subjects who supervise the platforms used by minors, as they maintain an 'exclusive relationship' with social media. The latter, as creators of digital platforms, according to risk proximity, are the most suitable to guarantee children's safety; therefore, they should be able to adopt such technology, without human intervention, to respect the specific vulnerabilities of the individuals. The digital environment, which has become a substantial part of people's social lives, should be made ethically and legally sustainable while balancing benefits and risks is justified under the values which represent the legal status of the minor.

4 Protecting Children in the Digital Age

While accessing Web resources, children impulsively provide their data. In this respect, Article 8 of the GDPR affirms, in relation to the offer of information society services directly to a child, that the processing of the personal data of a child shall be lawful where the child is at least 16 years old and has given their consent. However, Member States' national law may provide a lower age, though not below 13 years. Even below this threshold, the minor is not precluded from accessing those services by providing data, although the consent by the holder of parental responsibility for the child is still necessary.

Article 8 of the GDPR protects the interest of the child to develop their personality within social networks when the consent is provided; otherwise, it needs to be complemented by rules guaranteeing that free consent is given, at least for defending children from potential harm to their maturity level (vulnerability), which may isolate their identity.

If consent, as a particular aspect of private law, shall undoubtedly safeguard self-determination and autonomy in young children, at the same time it seems to be insufficient for the protection of their structural vulnerability. While they are not aware of the risks of social networks and the rights which have been recognised to them accordingly to their maturity level; especially in the context of the double spirit of the GDPR, which on the one hand aims at protecting the fundamental rights of the individual, on the other hand, enhances the free movement of personal data that "shall be neither restricted nor prohibited for reasons connected with the protection of natural persons with regard to the processing of personal data" (Article 1(3) GDPR).

To pursue these purposes of data protection, children's education plays a fundamental role, especially in the first years of life, while parents raise their children under a 'communion of life' (i.e., familial relationship), which follows the interaction between actors and interests. In the same communion of life, the digital environment also is included, in which places and people are connected, while parental education aims at making the child establish social relations in all its dimensions (physical, digital, etc.). The complexity of social life must be taken into account as a whole, by providing to the children the values of legality and civil life so that they do not think about 'free zones', outside the communion of life, where they could abdicate to their own identity.

For digital technologies, therefore, it is no longer sufficient to educate children within the usage of digital tools (the current generation does not need it), rather knowing how to behave in this digital dimension, not as a parallel world but as the propagation of individual social life and expression of personal identity, which shall remain unified.

This complexity of childhood education, beneath the principle of value objectivity, is quite singular because when the minor accesses the web they turn into the author of its education, not at the 'communional' level, as it happens in the familial relationship, but in the sterility of relations whose interests receive a mechanical satisfaction, without filters, inducing different interests. This leads to an individualistic and egoistic logic, in which there is not any sentimental and long-lasting aspect, typical of interpersonal relationships, but pure emotion.

Succinctly, parental education should be supported by technologies, which shall be designed in accordance with the law, as well as regulated, on a private basis, by those who develop and control them in a way that aims at being respectful of the vulnerabilities of individuals, especially users under the age of 18.

5 Child Vulnerability in the Digital World

The dictatorship of digital technologies seeks to move minor users from the 'communional' dimension. Parents, school, sports clubs, or religious associations play an educative role based on the interest of the child, either considered in the abstract (interest of *the* minor) or the context (interest of *a* certain minor), as in the case of parents who have to monitor the educational path about the rise of new interests as the child grows.

On the one hand, the child feeds the Net, by providing to the Web preferences respondent to their interests, which satisfy it. On the other hand, clicks, information, and data which are provided by the minor through the Web are governed by algorithms, while the child is satisfied and also conducted to targeted paths profiling, insofar that many 'filter bubbles' (Pariser 2011) condemn children to isolation and aphonia.

The main outcome is the rise of a digital identity in which the Network can educate and control the minor, by choosing the information that is linked to the identities of the algorithmic-made profiles.

The Net feeds the subject only with fleeting emotions and not with long-lasting feelings, by leaving the person to the pleasure of satisfying its desires and then to auto-isolate into the digital identity. However, it is necessary to ensure that personalities are formed in environments in which sensations do not cover feelings, especially where there are children, as their sociality is not reduced to sterile relationships detached from 'communional' aspects.

The feature of every 'filter bubble', regardless of its specific aims, is the non-discursive and critical participation of the person, who has become an unconscious recipient of the 'single thought' and output (in terms of information) of the algorithm. Those technologies deeply affect not only the relational dimension, which is increasingly drying up, but also the capacity for discernment of the subject to pursue its interests, thus distinguishing between choices that are good to them and vice versa.

The vulnerability of the minor, who must enjoy the protection of his parents, is exposed to serious risks, especially when the usage of technologies and smart devices starts becoming spasmodic under their exclusive control, outside the sovereignty of the adult.

Finding the necessary protection in the vulnerable situation requires a strict regulatory framework and dedicated remedies both of hard and soft law: preventive and repressive remedies have to be implemented not only from outside the digital environment but also within it, mainly to counterbalance the deviation of the technology that takes advantage of the user's fragility (in a broad sense). Therefore, the power of the algorithm affects real-world identity and creates many digital identities under the hidden interests of the Web and often aimed at achieving unimaginable economic benefits. Such identities are captured, hidden, as far as the values that constitute the real-world identity are dismissed, whether left to the uncritical and non-relational logic of the algorithm, by leaving the person in total isolation.

The level of risks that affects data subjects directly shows the inadequacy of public control or mandatory rules, which aims at governing the digital environment from the outside, also due to their a-territoriality. Otherwise, the private autonomy should be recalled, thus making private law assume a regulatory function, by entrusting network operators on self-regulating the platforms, as well as establishing control and reaction tools that can be activated by the users themselves, as it happened in other areas (e.g., antitrust law) in which the private enforcement has been used.

In essence, the digital ecosystem should be both legally and ethically compliant, even when developing or designing the algorithm, to adapt it to the specific conditions of the child user. However, a clear distinction should be made between the user who has the capacity for discernment and who has not, thus integrating all the conditions for the affirmation of the relationship between humans and technology, in which the latter is designed to serve mankind and not vice versa.

6 Cyberbullying and Digital Contents

The regulation of the digital environment left to both public and private law can be also found in recent regulatory acts concerning the activity involving minors in the network; such activities, among other outcomes, may lead them to 'bubbles' which make them victims of acts of bullying, as well as recipients of misleading information (fake news) which also contains hate speech and verbal violence.

An example is the Italian Law of 29 May 2017, No. 71, concerning certain aspects for the safeguarding of minors and the prevention and tackling of cyberbullying.

In essence, both preventive and remedial tools either of public or private law have been taken, by giving, especially to the latter, a proper regulatory function. Accordingly, the effectiveness of protection requires forms of private regulatory schemes in the digital environment, by implementing a sort of 'social responsibility' of digital platforms, following widespread control measures of the users.

However, cyberbullying refers to a complex phenomenon, which often takes place in the 'peer group'. In this respect, cyberbullying is meant as a series of actions—including also the unlawful processing of personal data of a child—carried out in the digital environment, which is intentionally and predominantly aimed at creating a situation of isolation towards the minors by carrying out serious abuses, damaging activities or ridiculing them. Social networks are a 'breeding ground' for the diffusion of such phenomena: the child, by considering also the above-mentioned 'non-communional' dynamics, provides information relating themselves to the platforms which may be suitable to be abusively processed; on the other hand, those who intend to use these data attempting to the individual dignity could find in the online anonymity the place where to build up an identity different from real-world identity, thus performing what it is not permitted in the latter environment.

In the development of tools to prevent and fight cyberbullying, it is necessary to entrust the institutions traditionally deputed to support parental education with specific educational activities, such as courses, for enforcing a culture of rights. In this respect, a technical board has been set up at the Presidency of the Council of Ministers, in order to develop an action plan that also may involve social services and schools; at the same time, the Italian Government should run periodic information campaigns for preventing and increasing awareness of cyberbullying through mass media communication, social media, and private actors. The plan needs to be implemented by a code of co-regulation to which all social networking service providers and the whole internet providers must opt-in; the code shall also create a monitoring committee in order to ensure the effectiveness of protection provided by the law.

Concerning preventive measures, schools will play a crucial role—following specific guidelines settled down by the Ministries for Education and University—as a place of education (also in a broad sense) for training students over legality and culture of rights, having particular regard for the usage of digital technologies, while increasing children awareness over the deviation of algorithms. School education and training activities should be carried out not only with public bodies, such as social services, but also with private entities (associations, local youth centres), including also the educational role played by the same students according to the peer education system. The school, with its training tools, has to support child victims and re-educate the perpetrators of such forms of violence.

Repressive remedies are also left to several public and private bodies, according to the principle of proximity. First of all, a 14-year-old child is allowed to provide personal data, as well as having the right to erasure—from the data controller, the website manager, or the social media—or restrict the contents which are detrimental to their dignity. If this is not possible or is not being enforced, the request may be addressed to the Data Protection Authority (DPA). Each parent or the adult who exercises parental responsibility should undertake actions to this end since they are responsible for the education and care of the child. Even the school manager, when notices acts of cyberbullying, shall immediately notify the subjects exercising parental responsibility or the guardians of the minors involved and adopt appropriate educational actions.

In short, the effectiveness of protection relies on a network of subjects and regulatory measures that are needed, first and foremost, to rule the web society in accordance with respect for human dignity.

7 The EU's Audiovisual Media Services Directive

Private regulation is explicitly recognised now in the Audiovisual Media Services Directive (Dir. EU 2018/1808) to ensure the effectiveness of protection of a person under the age of 18. In light of the massive use of video-sharing platforms by minors and, more broadly, information spread also through social media, EU Institutions have extended to them the restrictions already provided for audiovisual media services due to their products being potentially harmful to health and/or content inciting to violence and hatred (hate speech).

The Directive leaves the media service providers to adopt appropriate security measures, information disclosure, and control, aimed at protecting the child user from harm to their personality development, insofar that any further use is to be considered abusive. In this sense, the Directive supports the activation of such measures (e.g., for age verification) aimed at protecting the personal data of the child. In particular, data shall not be further processed for commercial purposes, such as profiling the minor by moving them to the 'filter bubbles'. However, profiling is not forbidden in general, but it requires specific safeguards, especially in terms of information to be provided to the data subject, as well as the data controller has obtained valid consent.

The call for code of conduct, as a self-regulatory or co-regulatory measure, is also encouraged by extending means to the digital environment which have already been adopted for offline communication channels.

Once more, the regulatory measures rely on soft law, although they lack the precision that characterises the hard law, but finds their strength in the preventive control of the rules adopted according to the protection purposes led by the legislator. By considering the isolation in which the child users find themselves on the network, such control from the outside, whether public or private, may certainly clash with boundless and impenetrable areas; thus, it is essential that the communication service provider resorts to social responsibility, acting as guardian of the child who spends time with its environments.

The Directive mentions the criteria for the platform regulation, whether led by the State or by other actors, which consists in balancing the fundamental rights, on a case-by-case basis. This can be the right to respect for private and family life, the right to data protection, the freedom of expression and information, the freedom to conduct a business, the prohibition of discrimination, and the rights of the minor. The principle of proportionality must guide the choice of which measures should be adopted, taking into account the riskiness of the information to forbid access by minors to contents that may act against their physical, mental, and moral development. Among the measures which could be considered appropriate, there is also the adoption by social media platforms of tools for user age verification.

Access to control systems shall be an effective preventive mechanism to ensure children's safety. A recent Italian case involving the death due to asphyxiation of a ten-year-old girl, who was registered on various social networks, had convinced the Italian DPA to act against the social media platform TikTok, suspecting that the young girl may have been exposed to harmful content posted online. Among different charges, the aforementioned digital platform would have failed to take adequate measures for age verification, especially for minors. Therefore, the DPA has prohibited, on a provisional basis, the Chinese App TikTok the further processing of personal data of users throughout the national territory for which there is no absolute certainty of age, in compliance with the provisions related to the age requirement. Accordingly, the same authority also asked Facebook—who also owns Instagram—to provide more details on the methods of registration to its social platforms and how the user's age verification procedure would ensure compliance with the minimum age for the subscription.

8 Concluding Remarks

The problems outlined in this article aimed to clarify that children should be protected primarily from the risk of isolation within their digital social life, which increasingly expands in the Net with boundless space. This constitutes a serious threat to individual dignity, as the most prominent value of human being, by making children lose the 'communional' dimension of their relationship, as well as providing an uncritical, relativistic, and dumb solipsism, which does not feel the need for any rules of peaceful coexistence.

The relationship between humans and digital technologies should be defined under the conformation of the latter directly by those who design and develop the former, in accordance with rules that make digital tools compatible with human dignity, as well as under the constant control of their users.

In this respect, the instrument of private regulatory law can be foreseen as an indispensable prerequisite for the effectiveness of the protection pursued by the hard law.

References

Andreola E (2019) Minori e incapaci in Internet. Edizioni Scientifiche Italiane, Napoli
Andreoli V (2021) La famiglia digitale. Come la tecnologia ci sta cambiando, Solferino, Milano

Bianca M (2019) La filter bubble e il problema dell'identità digitale. MediaLaws—Rivista di diritto dei media (2):39–53
Di Sabato D (2020) Diritto e new economy. Edizioni Scientifiche Italiane, Napoli
Donati F (2019) La tutela dei minori nella direttiva 2018/1808. Media Laws (1):60–72
Livingstone S (2018) Children: a special case for privacy? Intermedia 46(2):18–23
Lupton D, Williamson B (2017) The datafied child: the dataveillance of children and implications for their rights. New Media Soc 19(5):780–794
Pariser E (2011) The filter bubble: what the internet is hiding from you. Penguin Books Limited, New York
Perlingieri C (2016) La tutela dei minori di età nei social networks. Rass dir civ (4):1324–1340
Pitruzzella G (2018) La libertà di informazione nell'era di Internet. Media Laws (1):19–47
Senigaglia R (ed) (2019) Autodeterminazione e minore età. Itinerari di diritto minorile. Pacini Giuridica, Pisa
Senigaglia R (2020) Minore età e contratto. Contributo alla teoria della capacità. Giappichelli, Torino
Stoilova M, Livingstone S, Nandagir R (2020) Digital by default: children's capacity to understand and manage online data and privacy. Media Commun 8(4):197–207
Viglione F (2020) Riflessioni sui rimedi civilistici all'*hate speech*. Riv dir civ (4):775–795
Zeno-Zencovich V (2018) Dati, grandi dati, dati granulari e la nuova epistemologia del giurista. Media Laws (2):32–38
Zoppini A (2020) Il diritto privato e i suoi confini. Il Mulino, Bologna

Perspectives

Personal Data, Non-personal Data, Anonymised Data, Pseudonymised Data, De-identified Data

Claudia Irti

1 Personal Data, Non-personal Data

'Personal data' is the *material* scope of *data protection law*: only if the data subjected to processing is 'personal data', the General Data Protection Regulation—Regulation (UE) 2016/679 (GDPR)—will apply. 'Data' that is not personal data—and that we will call *non-personal data*—can be freely processed within the legal framework of the Regulation (UE) 2018/1807, the Regulation of the European Parliament and of the Council of 14 November 2018 on a framework for the free movement of non-personal data in the European Union.

So, distinguishing exactly what is meant by 'personal data' and by 'non-personal data', it is necessary to define the scope of application of the respective discipline and, especially, to establish if the entity processing data is subject to the various obligations that the GDPR imposes on data controllers. In doing so, it is necessary to start from the normative definitions.

Under Article 2(1) of the GDPR, 'personal data' means 'any information *relating* to an *identified* or *identifiable* natural person ("data subject"); an identifiable natural person is one who can be identified, directly or indirectly, in particular by reference to an identifier such as a name, an identification number, location data, an online identifier, or to one or more factors specific to the physical, physiological, genetic, mental, economic, cultural, or social identity of that natural person'; while, under Article 3(1) of the Regulation (UE) 2018/1807, what we call 'non-personal data' 'means data other than personal data as defined in point (1) of Article 4 of Regulation (EU) 2016/679'.

Our goal, therefore, seems to depend on what constitutes personal data, notwithstanding, is not so easy to circumscribe what it means because the different elements

C. Irti (✉)
Ca' Foscari University of Venice, Venice, Italy
e-mail: claudia.irti@unive.it

R. Senigaglia et al. (eds.), *Privacy and Data Protection in Software Services*, Services and Business Process Reengineering,
https://doi.org/10.1007/978-981-16-3049-1_5

of the definition—'any information', 'relating to', 'an identified or identifiable natural person'—is open to different interpretations that return a dynamic vision of it.

Starting with the normative definition, we learn that personal data is information about a natural person (not a legal person); it can take any form and be alphabetic, numeric, video or images; it includes both objective information (name, identification numbers, etc.) and subjective information (opinions, evaluations, etc.); the relevant element is that this information describes something about a subject that has value and meaning. Insignificant information, which has no meaning, should not be considered personal data, but new technologies have changed the way of attributing value to information because through them it is possible to collect, measure and analyse a lot of apparently 'insignificant' heterogeneous information that, reconnected to a person, are able to produce 'value'.

It is therefore essential to dwell on the concept of 'identifiability'.

According to Recital 26 GDPR, '(…) [t]o determine whether a natural person is *identifiable*, account should be taken of all the means that are reasonably likely to be used, such as detection, by the controller or another person, to identify the natural person directly or indirectly. To determine whether the means are reasonably likely to be used to identify the natural person, account should be taken of all objective factors, such as the cost and time required for identification, taking into account the technology available at the time of processing and technological developments'.

The Recital refers to the 'criteria' of the *reasonable probability of identification.*

Before the GDPR replaced Dir. 95/46/EEC, this 'criteria' was used to decide a case—the *Breyer case* (Case C-582/14)—submitted to the European Court of Justice, where the Court was asked to decide whether a dynamic IP address should be considered personal data, and the conclusion was that a dynamic IP address should be considered personal data. In this case, the Court expressly stated, for the first time, that information that allows the identification of a person does not need to be in the hands of a single individual, and to determine whether a person is identifiable, 'consideration should be given to the totality of the means likely reasonably to be used by the controller or others to identify the person'. At the same time, the Court reiterates that the risk of identification appears, in reality, to be insignificant if the identification of the data subject was prohibited by law or practically impossible on the account of the fact that it requires a disproportionate effort in terms of time, cost and man-power. In essence, the Court, as well as for the GDPR, admits that there can be a remaining risk of identification even in relation to 'anonymous' data: if the risk is limited, data can be treated as non-personal data, and this even though identification cannot be excluded with absolute certainty. The scope of personal data should be determined by assessing whether there is a means reasonably likely to be used to identify individuals and not merely a theoretical possibility of identification. This is what has been called *the risk-based approach.*

As anticipated, Recital 26 states that a 'reasonable' investment of time and financial resources should be considered in determining whether a particular individual can be identified, but it is not obvious what standard of reasonableness should be applied because what is 'reasonable' depends, on subjective and objective evaluations; it depends on the context.

The same Recital 26 further requires that consideration be given to the technological tools used for personal identification that are not only currently available but also under development during the period for which the data will be processed. This is the main point because the 'criteria' of the *reasonable probability of identification* more and more appears intrinsically dependent on the 'technological context' in which the data is introduced, and the same consideration of the data as 'personal', consequently, turns out to be *dynamic*: the same set of data may not be identifiable at the beginning of the processing or from the point of view of the subject in charge of the processing because of the tools at their disposal, but they may become so later when the tools change or may have always been so from the perspective of another subject.

That the context factor is, in more general terms, of absolute importance in defining the concept of personal data has been well highlighted in a study commissioned from the University of Sheffield by the UK Information Commissioner, in which—through an empirical investigation of how the term 'personal data' has been interpreted and applied by data protection authorities in different European jurisdictions—it has been shown that the distinguishing factor lies in the use of the central criterion of relevance/irrelevance of 'context'. Countries adopting the 'context irrelevance' criterion suggest, explicitly or implicitly, that a list of data can be drawn up that are always (and/or never) personal data; context is not considered a crucial factor in determining whether data should be classified as 'personal'. Countries that, instead, adopt the 'context relevance' criterion classify (almost) all data as 'sometimes' capable of being qualified as personal data, meaning that all data could be qualified as personal data 'under the right circumstances'. As a result, these Countries believe that it is not possible to draw up a definitive list of data that will always (or never) constitute 'personal data'.

Another relevant issue concerns the possibility of combining different datasets, which are of the availability of different entities: an entity may have access to a dataset that at face value is anonymous but might then, purposefully or not, subsequently gain access to a dataset containing information that enables re-identification.

The fact that it is not possible to know what datasets a given controller has access to or may have access to in the future, as well as the fact that in the light of new technical possibilities seemingly 'insignificant data' may be aggregated to re-identify an individual, determining what the actual risk of re-identification, it seems very difficult.

It should also be considered that in many cases, the purpose pursued by the data controller is only to identify the data subjects, for example, in the case of personalised advertising: in this case, to claim that people are not identifiable, when the purpose of the processing is only to identify them, would be a contradiction in terms.

All these considerations make it clear that the most realistic way to approach the problem of distinguishing what it is meant by 'personal data' and by 'non-personal data', is to recognise the *dynamic* nature of the data: assuming that anonymous data become personal data as soon as the *linkability* becomes possible, it must be considered a controller's duty to monitor if and when this can happen and, consequently, to adopt technical and organisational measures to protect the individual in due time.

2 Anonymised Data

Although the European data protection framework recognises two categories of data—personal data and non-personal data—there is a third relevant category that is represented by data that were once personal but are no longer so because they have undergone processing that has led to anonymisation: this is what we call anonymised data.

Anonymisation is a form of data processing aimed at de-identification. Generally speaking, de-identification is a process to remove or obscure any personally identifiable information from individual records in a way that minimises the risk of inadvertent disclosure of individuals' identities and information about them. Data anonymisation is the process of data de-identification that produces data in which individual records cannot be linked back to the original because they do not include the translation variables necessary to do so. Although it is not possible to completely remove the risk of disclosure, this process is to be considered successful—from a techno-scientific point of view—when there is no reasonable basis to believe that the information remaining in the records can be used to identify an individual record.

The Article 29 Working Party (the 'WP29' which is now the European Data Protection Board, 'EDPB') has addressed the issue of anonymisation in several documents, stating that 'anonymised data' is data 'that previously referred to an identifiable person, but where such identification is no longer possible' and the subsequently qualifying anonymisation as 'a technique applied to personal data to achieve irreversible de-identification' in the meaning that 'the outcome of anonymisation as a technique applied to personal data should be, in the current state of technology, as permanent as erasure, i.e. making it impossible to process personal data'.

In its 2014 Opinion 05/2014 on Anonymisation Techniques, the WP29 took a very strict approach compared to the 'risk-based approach' taken by the GDPR (Recital 26) and the European Court of Justice (in the *Breyer case*). Although the WP29's opinions have no legal value—the ECJ never mentions them—they may nevertheless have some influence on how data law develops in case law.

The WP29 considers that three criteria ought to be considered to determine whether de-identification has occurred, namely, if (i) it is still possible to single out an individual; (ii) it is still possible to link records relating to an individual; and (iii) whether information concerning an individual can still be inferred: 'Where the answer to these three questions is negative, data can be considered anonymous'. In fact, the WP29 seems to consider that no amount of risk can be tolerated. A position that has been judged to be 'idealistic and impractical'.

Actually, it is widely admitted among specialists—a view substantiated by a range of practice studies—that it is relatively easy to identify an individual through the combination of various *anonymised* datasets.

A study conducted by the Cambridge Institute of Technology (MIT), published in the journal Science in 2014, confirms that through the extraction and aggregation of non-identifying data, it is possible to trace a person's identity, de-anonymising them. The study was based on the analysis of credit card transactions made over the course of three months, an analysis from which it was possible to track the spending of 1.1 million people in 10,000 stores in a single country. The bank did not provide names, credit card numbers, store addresses or even the exact times of the transactions but only *metadata*: the amounts spent, the type of store (restaurant, gym, grocery store, etc.) and a code that represents each person. Because each individual's spending pattern is unique, the data detected very high 'uniqueness' making it suitable for what has been called a 'correlation attack'. In order to trace the identity of each individual, it was sufficient to relate the metadata to information about the person from external sources.

Everyone is willing to share daily a huge amount of personal data in different interactive systems. From all these data, collected in databases and exchanged and combined for different purposes (advertising, public security and so on), it is possible to collect such a large number of identification elements that it is abstractly possible to reconnect each data to a person. Pretending that data anonymisation can be 'as permanent as erasure', it is simply utopic in an increasingly digitised environment.

The 'absolute approach' of the WP29 would make it necessary to abolish the concepts of anonymous information and to reconsider the entire EU system of personal data protection, based on the distinction between 'personal data' and 'non-personal data'. But, as has been noted, 'anonymizing data, even with a small residual risk of re-identification, may be a more effective means of protecting the rights and interests of data subjects than leaving this data in its initial state' (Finck and Pallas 2020), applying duties to controllers and rights to data subjects that they are however unlikely to enforce.

Instead, the risk-based approach that the GDPR embraces seems to be a useful criteria to determine whether data qualifies as personal data; accepting that there always remains a residual risk of identification even where data is anonymised, it seems important to address risk with a two-step approach, as GDPR Recital 26 seems to do: it first requires a forecasting of the future, and second that decisions are made on the basis of that forecast. Using these criteria, it should be possible to identify different levels (and risks) of re-identification in different areas, based on which organise different models of privacy by design.

3 Pseudonymised Data, De-identified Data

The above considerations allow us to better understand why the GDPR speaks of data that are processed to reduce their *linkability* with individuals as 'pseudonymous data' rather than 'anonymous data'.

The 'pseudonymisation'—as explained in Article 4(5) of the GDPR—is a 'processing of personal data in such a manner that the personal data can no longer be attributed to a specific data subject without the use of additional information, provided that such additional information is kept separately and is subject to technical and organisational measures to ensure that the personal data are not attributed to an identified or identifiable natural person'. A definition that appears for the first time in the GDPR, whereas it was not contemplated in the Directive 95/46/EEC.

How it was observed, otherwise, that 'the GDPR does not merely describe a technique or a process—in fact, it does not specify at all what techniques should be used, other than stating that a "process" must be applied. GDPR pseudonymisation requires not just a process but an ultimate "success state", in which the data cannot be attributed to an individual without the use of additional information. Even this additional information is addressed within the definition, as it must be subject to "technical and organisational measures" to prevent reattribution. Thus, the data must not only be modified so that they are not directly identifiable, but they must also be protected against re-identification'.

This is why the WP29's finding that 'pseudonymisation is not a method of anonymisation. It merely reduces the linkability of a dataset with the original identity of a data subject, and is accordingly a useful security measure'.

As the definition makes clear, 'pseudonymised data' remains 'personal data' which are within the scope of the GDPR, and the data subject rights set out in Articles 15–20 still apply.

Recital 28 of the GDPR suggests the application of pseudonymisation to personal data as a useful tool that 'can reduce risks to data subjects and help controllers and processors comply with their data protection obligations': it is considered a technical measure that can be adopted by the data controller to comply with the security obligations imposed by data protection.

More exactly, the 'pseudonymisation processing' of personal data seems to be a specific form of data protection that must make it impossible to attribute personal data to a specific person except with the use of additional information, which must therefore be stored separately. It is thus the additional identifiable information, held separately from the pseudonymised data, which must be protected. An explanation of this definition of pseudonymised data can be found in Recital 29 GDPR: 'In order to create incentives to apply pseudonymisation when processing personal data, measures of pseudonymisation should, whilst allowing general analysis, be possible within the same controller when that controller has taken technical and organisational measures necessary to ensure, for the processing concerned, that this Regulation is implemented, and that additional information for attributing the personal data to a specific data subject is kept separately. The controller processing the personal data should indicate the authorised persons within the same controller'. Reading this provision, it is clear that protection is required for pseudonymised data against what might be called 'internal' identification risk: the danger that additional information retained by the data controller or a known third party might be matched to pseudonymised data for re-identification.

The controller may also delete information with which it would be possible to re-identify the data subject.

In this case, Article 11 of the GDPR limits some of the data controllers' obligations, providing that data that has been de-identified may be exempt from certain data subject rights, such as access, correction, deletion and portability requests as long as controllers can demonstrate that they cannot identify the data subject. This article exactly provides: (1) 'If the purposes for which a controller processes personal data do not or do no longer require the identification of a data subject by the controller, the controller shall not be obliged to maintain, acquire or process additional information in order to identify the data subject for the sole purpose of complying with this Regulation'; (2) 'Where, in cases referred to in paragraph 1 of this Article, the controller is able to demonstrate that it is not in a position to identify the data subject, the controller shall inform the data subject accordingly, if possible. In such cases, Articles 15 to 20 shall not apply except where the data subject, for the purpose of exercising his or her rights under those articles, provides additional information enabling his or her identification'. Therefore, the controller who de-identifies the personal data for processing, and no longer has any interest in re-identification, has no obligation to maintain that information that would be necessary for re-identification for the sole purpose of complying with the GDPR; in this case, the controller must, where possible, inform the data subject in advance that they will no longer be able to re-identify. Only if the data subject is, itself, able to provide personal identifying keys, then the controller is required to recognise the data subject's rights provided by Articles 15 to 20. The regulatory provision encourages the data controller to use processing techniques that reduce the possibility of re-identification of data subjects by proportionally reducing the obligations imposed on him.

We have so begun to talk about *de-identified data.*

This is an expression that, in general terms, refers to data that has been subjected to a process that removes or obscures any personally identifiable information from individual records in a way that minimises the risk of inadvertent disclosure of individuals' identities and information about them.

In this sense, pseudonymised data are also de-identified data.

Some legal scholars, however, use this expression to indicate data that, on a hypothetical scale of identifiability, are one step above pseudonymised data.

In fact, while pseudonymised data are those treated in such a way that 'personal data can no longer be attributed to a specific person without the use of additional information, provided that such additional information is kept separately and subject to technical and organizational measures to ensure that such personal data are not attributed to an identified or identifiable natural person' (Article 4(5) GDPR), 'de-identified' data would be those in which the data controller is no longer able to re-identify the data subject unless the additional information necessary for re-identification is provided by the data subject himself, who is the only one who has it. This differentiation is introduced by Article 11(2) of the GDPR, which for this reason can be considered as the provision that introduces the recognition of multiple levels of identifiability.

The standard appears to refer to additional information that is held by the data controller (Article 11(1)) or the data subject themself (Article 11(2)); according to the WP29 and the European Court of Justice (*Breyer case*), instead, it is also necessary to refer to additional information that may be held by third parties. In practice, however, data controllers can often only assess the re-identification risk of their own activities, as it is not clear which 'third parties' should be considered.

In any case, if the process of data de-identification is intended to reduce the risk of data subject identifiability, in applying the *risk-based approach* we should say aloud that personal data that has been de-identified may fall within the scope of the GDPR depending on how difficult it is to re-identify the data subject. A proportionate and context-dependent approach would take into account the range of organisational, legal and technological measures (including pseudonymisation) to determine whether the data is considered (or not) identifiable and consequently whether the controller is subject (or not) to obligations under the GDPR.

Depending on the case, the 'means that are reasonable to use for identification' test may be limited to identifying data through additional known information that is held separately—by the data controller, the data subject or third parties—or may also include, more indirect methods of identification, such as singling out individuals and determining their identity using other multiple sources of information.

References

Article 29 Working Party (2007) Opinion 04/2007 on the concept of personal data. https://ec.europa.eu/justice/article-29/documentation/opinion-recommendation/files/2007/wp136_en.pdf

Article 29 Working Party (2014) Opinion 05/2014 on anonymisation techniques. https://ec.europa.eu/justice/article-29/documentation/opinion-recommendation/files/2014/wp216_en.pdf

Bohannon J (2015) Privacy. Credit card study blows holes in anonymity. Science 347:468–468

Booth S, Jenkins R, Moxon D, Semmens N, Spencer C, Taylor M, Townend D (2004) What are 'personal data'? A study conducted for the UK Information Commissioner, University of Sheffield. https://www.frareg.com/cms/wp-content/uploads/personal_data.pdf

Burgin M (2010) Theory of information. Fundamentality, diversity and unification. World Scientific Publishing, Singapore

Ducato R (2016) La crisi della definizione di dato personale nell'era del web 3.0. Una lettura civilistica in chiave comparata. In: Cortese F, Tomasi M (eds) Il diritto e le definizioni. Edizioni Scientifiche Italiane, Napoli, pp 145–178

Finck M, Pallas F (2020) They who must not be Identified—distinguishing Personal from Non-Personal Data under the GDPR. Int Data Priv Law 10(1):11–36

Floridi L (2009) Philosophical conceptions of information. In: Sommaruga G (ed) Formal theories of information: from Shannon to semantic information theory and general concepts of information. Springer, Berlin, Heidelberg, pp 13–53

George D, Reutimann K, Tamò-Larrieux A (2019) GDPR bypass by design? Transient processing of data under the GDPR. Int Data Priv Law 9(4):285–298

Groos D, van Veen E-B (2020) Anonymised data and the rule of law. Eur Data Prot Law Rev 4(6):1–11

Irti C (2020) Dato personale, dato anonimo e crisi del modello normativo dell'identità. Ius civile (2):379–397

Mourby M, Mackey E, Elliot M, Gowans H, Wallace SE, Bell J, Smith H, Aidinlis S, Kaye J (2018) Are 'pseudonymised' data always personal data? Implications of the GDPR for administrative data research in the UK. Comput Law Secur Rev 34(2):222–233

Ohm P (2010) Broken promises of privacy: responding to the surprising failure of anonymization. UCL Law Rev 57:1701–1711

Pellecchia E (2020) Dati personali, anonimizzati, pseudonimizzati, de-identifcati: combinazioni possibili di livelli molteplici di identificabilità nel GDPR. Le Nuove Leggi Civili Commentate 2:360–373

Purtova N (2018) The law of everything. Broad concept of personal data and future of EU data protection law. Law Innov Technol 10(1):40–81

Stalla-Bourdillon S, Knight A (2017) Anonymous data v. personal data—a false debate: an EU perspective on anonymisation, pseudonymisation and personal data. Wis Int Law J 34:284–322

Zech H (2015) Information as property. J Intell Prop Inf Technol Electron Comp 6:192–197

Zech H (2016a) A legal framework for a data economy in the European Digital Single Market: rights to use data. J Intellect Prop Law Pract 11(6):460–470

Zech H (2016b) Data as a tradeable commodity. In: De Franceschi A (ed) European contract law and the digital single market. Intersentia, Cambridge, pp 51–81

Personal Data as Counter-Performance

Alberto De Franceschi

1 Introduction

In its recently published 'Strategy for Data', the European Commission mentions that the volume of data produced globally is estimated to grow from 33 zettabytes (a zettabyte is 10^{21}) in 2018 to 175 zettabytes in 2025. In particular, personal data-driven business models are experiencing an unprecedented growth.

From a systematic point of view, it is possible to distinguish between two main categories of personal data based business models: (i) those which do not foresee any monetary payments by the user, but rather the provision of data and the consent to their processing (this is the case, e.g., of Facebook and Google); (ii) those which provide a basic version without requiring for the access to it the payment of an amount of money, but at the same time offering a premium version against a monetary counter-performance (e.g., Dropbox and LinkedIn). Furthermore, it is possible to distinguish between two types of business models which provide a monetary counter-performance and an additional counter-performance consisting of personal data together with the consent to their processing: (i) those which foresee a monetary counter-performance and, in addition, a discount conditioned to the provision of personal data and the consent to their processing (see specific kinds of insurances as, e.g., the 'pay as you drive' models); (ii) those in which the consumer provides a monetary counter-performance and the trader processes his or her data for creating value (this is the case of most online traders, like Amazon, or intermediaries, like Airbnb).

In the data-driven business models, data trade and data protection issues are closely interconnected. Both aspects are subject of analysis in this contribution. Within the

A. De Franceschi (✉)
University of Ferrara, Ferrara, Italy
e-mail: alberto.defranceschi@unife.it

R. Senigaglia et al. (eds.), *Privacy and Data Protection in Software Services*,
Services and Business Process Reengineering,
https://doi.org/10.1007/978-981-16-3049-1_6

outlined framework, this paper focuses *inter alia* on the solution introduced by Directive 2019/770/EU on certain aspects concerning contracts for the supply of digital content and digital services, whose scope of application is limited to business to consumer contracts. Against this background, the paper starts from the analysis of the provisions on data protection and their interplay with the law of obligations. From this situation emerges a fragmented picture of the national and European rules regarding the 'payment with personal data'.

The interaction and the need for coordination with the implementing provisions of Directive 2011/83/EU on consumer rights (hereinafter: CRD) as well as with Directive 2005/29/EC on unfair commercial practices (hereinafter: UCPD) will be also subject of analysis.

2 The Development of the Notion of Personal Data and the Requirements for Their Processing

Article 3(1) of the Dir. 2019/770/EU (hereinafter: DCD) provides that the Directive shall apply where the trader supplies or undertakes to supply digital content or digital service to the consumer, and the consumer provides or undertakes to provide personal data to the trader, except where the personal data provided by the consumer are exclusively processed by the trader for the purpose of supplying the digital content or digital service in accordance with the Directive or for allowing the trader to comply with legal requirements to which the trader is subject, and the trader does not process those data for any other purpose.

According to Article 2(1)(8) DCD 'personal data' means personal data as defined in Article 4 of the Regulation 2016/679/EU (hereinafter: GDPR), which qualifies as personal data any information relating to an identified or identifiable natural person ('data subject'); an identifiable natural person is one who can be identified, directly or indirectly, in particular by reference to an identifier such as a name, an identification number, location data, an online identifier or to one or more factors specific to the physical, physiological, genetic, mental, economic, cultural or social identity of that natural person. Compared to the definition already contained in the Directive 1995/46/EC, the GDPR significantly expands the notion of personal data, including location data, an online identifier or one or more factors specific to the physical, physiological, genetic, mental, economic, cultural or social identity of that natural person.

Article 6 GDPR sets the requirements for the lawfulness of the processing of personal data. In particular, the provisions contained in Article 6(1)(a) and (b) GDPR are particularly relevant for the subject of this analysis, as they provide authorizations—by the data subject or by the law—to the processing of personal data.

As regards the consent to the processing of personal data, Article 6(1) GDPR provides that the processing is lawful only if at least one of the conditions listed in the same paragraph are fulfilled: "(a) the data subject has given consent to the processing of his or her personal data for one or more specific purposes; (b) processing is necessary for the performance of a contract to which the data subject is party or in order to take steps at the request of the data subject prior to entering into a contract". According to Article 6(1)(c)–(f) GDPR the authorisation can be given by the law when public interest makes this "necessary". In this regard, the requirement of 'necessity' shall be interpreted restrictively, in order not to excessively limit the innovative potential of the DCD.

The conditions for lawful consent are determined in Article 7 GDPR. In particular, Article 7(1) GDPR provides that, where processing is based on consent, the controller shall be able to demonstrate that the data subject has consented to the processing of his or her personal data. If the data subject's consent is given in the context of a written declaration, which also concerns other matters, according to Article 7(2) GDPR the request for consent shall be presented in a manner which is clearly distinguishable from the other matters, in an intelligible and easily accessible form, using clear and plain language. Any part of such a declaration, which constitutes an infringement of this Regulation, shall not be binding.

In this regard, Article 7(4) GDPR provides the conditions to determine if the consent has been given voluntarily or not. Indeed, according to that provision, when assessing whether consent is freely given, utmost account shall be taken of whether, *inter alia*, the performance of a contract, including the provision of a service, is conditional on consent to the processing of personal data that is not necessary for the performance of that contract. Therefore, the question is whether the consent the data subject has given to the processing of their personal data was extorted and therefore shall be considered ineffective. Such risk could actually exist only in the case of a monopolistic supplier.

Significantly stricter are the requirements contained in Article 8 GDPR regarding the conditions applicable to child's consent in relation to information society services: here it is expressly provided that where Article 6(1)(a) GDPR applies, in relation to the offer of information society services directly to a child, the processing of the personal data of a child shall be lawful where the child is at least 16 years old. Where the child is below the age of 16 years, such processing shall be lawful only if and to the extent that consent is given or authorised by the holder of parental responsibility for the child.

According to Article 7(3) GDPR, the data subject shall have the right to withdraw his or her consent at any time. The withdrawal of consent shall not affect the lawfulness of processing based on consent before its withdrawal. Prior to giving consent, the data subject shall be informed thereof. It shall be as easy to withdraw as to give consent. In any case, the indispensability of the consumer withdrawal right is already rooted in Article 8 of the Charter of Fundamental Rights of the European Union.

3 The Circulation of Personal Data from the Perspective of the Law of Obligations

3.1 Personal Data as Counter-Performance and the Notion of Price

Data have a growing monetary value and users are increasingly used to pay with them rather than with money: although, this does not mean that consumers are always conscious of the circumstance that when supplying personal data and agreeing to their processing they are indeed paying or providing a performance which has a value similar to money. In this regard, the innovation brought about by the DCD plays a crucial role as it generated a lively debate about the possibility to configure data as a counter-performance. Some members of the European Council were reluctant to include data as counter-performance into the Directive, referring to an opinion of the European Data Protection Supervisor (EDPS), whose argument was that "personal data cannot be compared to a price or money. Personal information is related to a fundamental right and cannot be considered as a commodity". A counterargument was that of course this should be admitted, given that the fundamental right to one's own image is already subject to commercialization. Furthermore, the EDPS stressed the argument according to which "there might be well a market for personal data, just like there is, tragically, a market for live human organs, but that does not mean that we can or should give that market the blessing of legislation. One cannot monetise and subject a fundamental right to a simple commercial transaction, even if it is the individual concerned by the data who is party to the transaction". Such argument was countered by the acknowledgement that every day a huge number of such contracts providing data as a counter-performance are concluded and nobody thinks about treating them as trade of live organs.

After intense discussion, the majority followed the European Commission's proposal and therefore the DCD's scope of application covers now also contracts in which the consumer provides to the supplier personal data as a counter-performance. The final version of the Directive nevertheless adopted a somehow 'softer' formulation (compared to that contained in the EU Commission's proposal), avoiding using the formulation—originally contained in Article 3 of the EU Commission's proposal—"the consumer (…) provides counter-performance other than money in the form of personal data or any other data (…)". Indeed, Article 3(1) DCD states that the Directive "shall also apply where the trader supplies or undertakes to supply digital content or a digital service to the consumer, and the consumer provides or undertakes to provide personal data to the trader". Recital 24 DCD further clarifies the point, taking into consideration the previous debate and stating that, while fully recognising that the protection of personal data is a fundamental right and that therefore personal data cannot be considered as a commodity, the Directive should ensure that consumers are, in the context of such business models, entitled to contractual remedies. This Directive should, therefore, apply to contracts where the trader supplies, or undertakes to supply, digital content or digital service to the consumer, and the consumer provides or undertakes to provide personal data. The personal data could be provided to the trader either at the time when the contract is concluded or at a later time, such as when the consumer gives consent for the trader to use any personal data that the consumer might upload or create with the use of the digital content or digital service. For example, the Directive should apply where the consumer opens a social media account and inserts the name and email address that are used for purposes other than solely supplying the digital content or digital service, or complying with legal requirements. The DCD shall equally apply where the consumer gives consent for any material that constitutes personal data, such as photographs or posts that the consumer uploads, to be processed by the trader for marketing purposes. Member States should however remain free to determine whether the requirements for the formation, existence and validity of a contract under national law are fulfilled.

Considering the mentioned evolution, it is particularly desirable that at the European level it comes to the approximation of the treatment of the contractual schemes 'digital contents and services for money' and 'digital contents and services for personal data'. For this reason, the notion of 'price', contained in Article 2(1)(7) DCD shall in the author's opinion be 'updated' taking into account the needs of the digital economy. Indeed, according to the aforementioned provision 'price' means "money or a digital representation of value that is due in exchange for the supply of digital content or a digital service", while it could now include "personal data which is due in exchange for the supply of digital content of a digital service".

A first although incomplete step in this direction is marked by Recital 23 DCD, which acknowledges that digital representations of value such as electronic vouchers or e-coupons are used by consumers to pay for different goods or services in the Digital Single Market. Such digital representations of value are becoming important in relation to the supply of digital content or digital services and should therefore be considered as a method of payment within the meaning of the Directive. According to the same Recital, digital representations of value should also be understood to include virtual currencies, to the extent that they are recognised by national law. As underlined in the same recital, differentiation depending on the methods of payment could indeed be a cause of discrimination and provide an unjustified incentive for businesses to move towards supplying digital content or a digital service against digital representations of value.

3.2 *The Role of the Consent to the Processing of Personal Data*

Given that the supply of personal data by the consumer is increasingly subject to performance, in such cases, the trader has above all interest to obtain from the consumer their consent to the processing of the personal data so that he or she will be able to profit from them. Indeed, the collection of personal data is useful for whoever receives the data as he or she is allowed to use those data. This can happen, for example, in the case of activation of targeted advertising or data resale. So the mere access to personal data is usually not of interest for the trader, as a subsequent data transfer to third parties is, without the consent of the data subject to the processing of personal data, in the majority of cases tendentially irrelevant (see Article 6 et seq. GDPR). Regarding the complex interplay between data trade and data protection, the DCD does not contain any precise indication, but rather merely refers to the GDPR, which states that any processing of personal data in connection with a contract falling within the scope of this Directive is lawful only if it conforms with the provisions of the GDPR itself relating to the legal grounds for the processing of personal data.

However, according to Article 3(2) DCD, the Directive shall not apply where the personal data provided by the consumer are exclusively processed by the trader for the purpose of supplying the digital content or digital service in accordance with the Directive itself or for allowing the trader to comply with legal requirements to which he or she is subject, and the trader does not process those data for any other purpose. Indeed, in such cases, the supply of personal data by the consumer cannot be considered as a counter-performance according to the general principles of the law of obligations. Nevertheless, the DCD finds again application and the supply of personal data (together with the consent to their processing) shall be considered as a counter-performance when the personal data are processed by the trader in a way that is not strictly necessary for the purpose of supplying the digital content or digital service in accordance with the DCD or for allowing the trader to comply with legal requirements.

Furthermore, personal data cannot be considered as a counter-performance where data processing is necessary for the execution of a contract to which the data subject is party or in order to take steps at the request of the data subject prior to entering into a contract (Article 6(1)(b) GDPR). However, the DCD finds again application if those personal data are used by the trader for commercial purposes. In this regard, a debate may arise also with regard to the determination of the concrete amount of data that would be necessary to the trader for the purpose of supplying the digital content or digital service. In particular, it is not clear if the amount of data which are necessary for the fulfilment of the contract shall be determined subjectively (from the perspective of the trader) or objectively. In any case, the requirement of the 'necessity' shall be interpreted restrictively.

3.3 Personal Data as Counter-Performance and Coordination with Directive 2011/83/EU on Consumer Rights and with Directive 2005/29/EC on Unfair Commercial Practices

Despite the increasing synergy between the regulation of the aspects of data trade and data protection, adequate coordination with the rules on information duties contained in Directive 2011/83/EU on consumer rights (hereinafter: CRD) is still missing. In particular, the DCD does not take adequately into consideration the pre-contractual information duties. It is remarkable that also the coeval Directive 2019/771/EU on the sale of goods does not deal with that aspect.

In this regard, the DCD merely states that instead of the provisions of the DCD on the trader's obligation to supply and on the consumer's remedies for failure to supply, the provisions of the CRD on obligations related to the delivery of goods and remedies in the event of the failure to deliver should apply. In addition, the provisions of Directive 2011/83/EU on, for example, the right of withdrawal and the nature of the contract under which those goods are supplied should also continue to apply to such tangible media and the digital content supplied on it (Recital 20 DCD); furthermore, the requirements of the contract should include those resulting from the pre-contractual information which, in accordance with Directive 2011/83/EU, forms an integral part of the contract. Those requirements could also be set out in a service level agreement, where, under the applicable national law, such type of agreement forms part of the contractual relationship between the consumer and the trader (Recital 42 DCD).

Nevertheless, as regards the aspect of remuneration of the supply of digital content or digital services the Directive on consumer rights shall find application. Before the consumer is bound by a distance or an off-premises contract or by a contract other than a distance or an off-premises contract, or any corresponding offer, the trader shall provide the consumer with a series of information in a clear and comprehensible manner, if that information is not already apparent from the context, as specifically regards the *total price* of the goods or services inclusive of taxes, or where the nature of the goods or services is such that the price cannot reasonably be calculated in advance, the manner in which the price is to be calculated, as well as, where applicable, all additional freight, delivery or postal charges or, where those charges cannot reasonably be calculated in advance, the fact that such additional charges may be payable (Articles 5(1)(c) and 6(1)(e) CRD).

However, the CRD does not contain any provisions dealing with the notion of price. As a consequence of that, some authors argue that from the context of the CRD it clearly emerges that the aforementioned notion can refer only to a sum of money. However, considering the above-outlined evolution of the social and economic framework, it would be adequate to interpret the notion of price in an evolutive way, including each counter-performance having economic value.

The webpages and cookies used for stimulating the consumer to authorise the collection and the processing of his or her data are often misleadingly labelled, e.g., with words like "registration" or "sign-in". From such indications the average consumer can most of the time not easily understand that what above conceals the proposal to conclude a contract which cannot be considered as gratuitous. Starting from the observation of the described situation and given the constantly large amount of complaints about 'Internet cost traps', Article 8(2) CRD provides specific 'formal requirements' for contracts concluded by electronic means. This rule has been proposed in order to enable consumers to be clearly and succinctly informed about all costs before entering into a binding contract. Therefore, Article 8(2) CRD lays down some specific formal requirements for contracts concluded by electronic means, which entail an obligation to pay. In particular, if a distance contract to be concluded by electronic means places the consumer under an obligation to pay, the trader shall make the consumer aware in a clear and prominent manner, and directly before the consumer places his order, of the information on the *essentialia negotii* provided for in Article 6(1)(a)(e)(o) and (p). The trader shall ensure that the consumer, when placing his order, explicitly acknowledges that the order implies an obligation to pay. If placing an order entails activating a button or a similar function, the button or similar function shall be labelled in an easily legible manner only with the words "order with obligation to pay" or a corresponding unambiguous formulation indicating that placing the order entails an obligation to pay the trader. If the trader has not complied with this subparagraph, the consumer shall not be bound by the contract or order. Even for the supply of digital content and digital services *versus* the supply of personal data and of the consent to their processing, the trader has to comply with the duties indicated in Article 8(2) CRD (in particular the duty to inform the consumer about the total price of goods or services: see Recital 50 and Article 6(1)(e) CRD). If the supply of personal data (together with the consent to their processing) can be considered as a 'payment', Article 8(2) CRD finds application: indeed, the aforementioned provision applies in all cases in which the trader does not clarify the non-gratuitousness of the contract.

Moreover, to the consequence of the non-bindingness of the consent to the processing of personal data, one could come on the basis of Article 7(2) GDPR, which provides that if the data subject's consent is given in the context of a written declaration which also concerns other matters, the request for consent shall be presented in a manner that is clearly distinguishable from the other matters, in an intelligible and easily accessible form, using clear and plain language. Any part of such a declaration that constitutes an infringement of this Regulation shall not be binding. From this, it can be derived that also the entire declaration can be considered as not binding if it (and not only parts of it) configure a breach of the GDPR.

This produces further consequences. Indeed, several online business models offer to consumers the conclusion of allegedly gratuitous 'framework contracts' for 0,00 Euro. However, such contracts are often not gratuitous, as the consumer 'pays' with his or her own personal data. When the trader does not comply with the provisions contained in Article 8(2) CRD, the aforementioned 'framework contract' is not binding for the consumer. The real goal behind such allegedly gratuitous contracts is to induce the consumer to stipulate one or more further contracts, which foresee a monetary counter-performance. This is, e.g., the case of health or professional network apps: the framework contract provides indeed a counter-performance consisting in 0,00 Euro (but, instead of money, personal data and the consent to their processing) and, in addition, the opportunity to conclude one or more additional contracts providing a counter-performance consisting in money (and personal data). By applying the aforementioned rules, if the framework contract is not binding for the consumer, also the following contracts, which are based on the framework contract, will be equally not binding for the consumer.

In case of data as a counter-performance, it is furthermore necessary to verify whether the trader violated the provisions of Directive 2005/29/EC on unfair commercial practices (hereinafter: UCPD). In particular, in the aforementioned cases of Internet cost traps, No. 21 of Annex I to the UCPD can apply. Indeed, that provision considers in all circumstances as unfair the commercial practice in which the trader describes a product as "gratis", "free", "without charge" or similar if the consumer has to pay anything other than the unavoidable cost of responding to the commercial practice and collecting or paying for delivery of the item. In this regard, particularly problematic is the case in which the trader offers to the consumer, in the same contract, a digital content or a digital service in exchange for an amount of money and an additional digital content or digital service, which is allegedly free, but in reality provided against an additional counter-performance consisting of personal data and the consent to their processing.

If the trader's behaviour does not fulfil any of the cases described in UCPD's Annex I, it will be necessary to check whether the existence of an unfair commercial practice can be assessed according to Articles 6, 7 (misleading practices) or Articles 8, 9 (aggressive practices), or, alternatively, according to the general prohibition of unfair commercial practices contained in Article 5 UCPD.

3.4 *The Withdrawal of Consent to the Processing of Personal Data: Effects on the Contract for the Supply of Digital Content and Digital Services and the Problematic Coordination with the Directive 2011/83/EU on Consumer Rights*

If the consumer allows the treatment of his or her personal data as counter-performance for the supply of digital content or digital services, according to Article 7(3) GDPR he or she shall have the right to withdraw his or her consent at any time. However, the consent withdrawal shall not affect the lawfulness of processing based on consent before its withdrawal. The indispensability of the right to withdraw the aforementioned consent already results from Article 8 of the Charter of Fundamental Rights of the European Union. A waiver of the right of withdrawal will be therefore ineffective.

This is suitable to cause a lack of stability of such contracts and has therefore a significant impact on the design of contracts for the supply or digital contract and/or digital services. Indeed, the possibility of withdrawing at any time the consent to the processing of personal data may cause significant problems for the case in which personal data are the counter-performance for the supply of digital content and/or digital services, also considering that the DCD does not provide anything in that regard.

There are a couple of possible solutions to the aforementioned question, even if it is not obvious that a withdrawal of the consent to the processing of personal data will automatically cause the contract termination. The contract providing the supply of digital content and/or digital services with personal data as counter-performance could be considered as a long-term agreement. As a consequence, the withdrawal of the consent to the processing of personal data could be considered—from the perspective of the provider—as the termination of the license for the utilization of digital content and/or digital service, and—from the consumer's perspective—the termination of the license for the use of personal data. Furthermore, it often happens that personal data belonging to a consumer are mixed with personal data belonging to other consumers for the purpose of data analysis, aggregation of data, data processing and production of other data or for transferring the data to third subjects. In the latter case, at the moment of contract termination, the results of the personal data analysis and processing are already outside the sphere of influence of the supplier of digital content and digital services, who initially received the personal data and the consent to their processing.

According to Article 7(3) GDPR, the withdrawal of the consent to the processing of personal data will not touch the legality of the already (until the moment of the consent withdrawal) undertaken data processing.

For the case of contract termination, Article 16(2) DCD generally provides that in respect of personal data of the consumer, the trader shall comply with the obligations applicable under the GDPR. In this regard, it was emphasised that considering the specific case of termination of a contract with data as a counter-performance it is almost impossible to carry out a full restitution, as the monetization of data through transfer or exploitation already took place before the moment of the termination. For this case, a compensation in money seems reasonable. In any case, the supplier will not be entitled to ask for compensation for contract termination. The consumer, however, can exercise the right to be forgotten according to Article 17 GDPR.

Furthermore, from a systematic point of view, the mandatory nature of the right of withdrawal set by Article 7(3) GDPR lets arise significant problems in coordination with the Directive 2011/83/EU on consumer rights. Indeed, Article 16 CRD ("Exceptions from the right of withdrawal") foresees that the EU Member States shall not provide for the right of withdrawal set out in Articles 9 to 15 CRD in respect of distance and off-premises contracts as regards the supply of digital content which is not supplied on a tangible medium if the performance has begun with the consumer's prior express consent and his acknowledgement that he thereby loses his right of withdrawal. By means of such provision the EU legislator tried to avoid possible abusive behaviours by the consumer, who, after the supply of the digital content may declare their withdrawal from the contract and ask the supplier for restitution of the amount paid, even if the consumer may have already durably saved the digital content in his or her digital environment. Evidently, the aforementioned provision was not conceived for encompassing also the case of personal data as a counter-performance. However, considering the legal framework as modified by the DCD, it would be reasonable to exclude the applicability of the above-mentioned exception to the right of withdrawal in the case of a contract providing personal data as a counter-performance, as in the latter case the exception contained in Article 16 CRD would contravene Article 7(3) GDPR and even more Article 8 of the Charter of Fundamental Rights of the European Union (see above, Section 2). Nevertheless, such (desirable) inapplicability is suitable to generate a disparity of treatment between 'paying with money' and 'paying with data', giving rise to a tension that shall be solved by the EU legislator in order to avoid unnecessary distortions.

4 Concluding Remarks

Data have an increasing economic value and consumers of digital content and/or digital services are now used to pay with data instead of or in addition to money. However, this does not mean that in such cases consumers are sufficiently aware that they are indeed paying with their own personal data. Indeed, in the majority of cases, consumers/users consider such supply of digital content/digital services in exchange for personal data as a free supply. Considering that evolution, it is therefore particularly desirable to come to an alignment of the treatment of contracts providing the supply of digital content/digital services against money and supply of digital content/digital services against personal data and the consent to their processing.

The EU Directive 2019/770 sets the first step in the right direction. Nevertheless, better coordination with the existing legal instruments—and especially with the GDPR and with the CRD—is needed. In particular, a clarification of the interaction between the consent to the processing of personal data and its consequences according to the law of obligations is necessary. The regime of information duties and the withdrawal of the consent to the processing of personal data also needs to be refined and better coordinated. Moreover, the notions of price and payment need to be interpreted in an evolutive way and their regulation in the coming legislative instruments shall be developed in order to adequately face the challenges of the digital economy.

References

Auer M (2019) Digitale Leistungen. Zeitschrift für die Gesamte Privatrechtswissenschaft 5(2):130–147

European Commission (2020) Communication from the Commission to the European Parliament, the Council, the European Economic and Social Committee and the Committee of the Regions. A European Strategy for Data, 19 February 2020, COM/2020/66 final. https://eur-lex.europa.eu/legal-content/EN/ALL/?uri=CELEX:52020DC0066

European Law Institute (2015) Statement on the European commission's proposed directive on the supply of digital content to consumers, COM (2015) 634 final. https://www.europeanlawinstitute.eu/fileadmin/user_upload/p_eli/Publications/ELI_Statement_on_DCD.pdf

Fries M (2020) Data as counter-performance in B2B contracts. In: Lohsse S, Schulze R, Staudenmayer D (eds) Data as counter-performance – Contract law 2.0?. Hart – Nomos, Oxford – Baden-Baden, pp 225–251

Graf von Westphalen F, Wendehorst C (2016) Hergabe personenbezogener Daten für digitale Inhalte–Gegenleistung, bereitzustellendes Material oder Zwangsbeitrag zum Datenbinnenmarkt? Betriebs Berater 37:2179–2187

Hacker P (2020) Datenprivatrecht. Mohr Siebeck, Tübingen

Hacker P (2020) Regulating the economic impact of data as counter-performance: from the illegality doctrine to the unfair contract terms directive. In: Lohsse S, Schulze R, Staudenmayer D (eds) Data as counter-performance – Contract law 2.0?. Hart – Nomos, Oxford – Baden-Baden, pp 47–76

Härting N (2016) Digital Goods und Datenschutz - Daten sparen oder monetarisieren? Die Reichweite des vom DinhRL-E erfassten Geschäftsmodelle. Comput Recht 32(11):735–740

Janeček V, Malgieri G (2020) Data extra commercium. In: Lohsse S, Schulze R, Staudenmayer D (eds) Data as counter-performance – Contract law 2.0?. Hart – Nomos, Oxford – Baden-Baden, pp 95–125

Langhanke C, Schmidt-Kessel M (2015) Consumer data as consideration. J Eur Consum Mark Law 4(6):218–222

Lapuente SC (2020) Termination of the contract for the supply of digital content and services, and availability of data: rights of retrieval, portability and erasure in EU law and practice. In: Lohsse S, Schulze R, Staudenmayer D (eds) Data as counter-performance – Contract law 2.0?. Hart – Nomos, Oxford – Baden-Baden, pp 163–191

Lohsse S, Schulze R, Staudenmayer D (2020) Data as counterperformance – contract law 2.0? An introduction. In: Lohsse S, Schulze R, Staudenmayer D (eds) Data as counter-performance – Contract law 2.0?. Hart – Nomos, Oxford – Baden-Baden, pp 9–21

Metzger A (2020) A market model for personal data: state of play under the new directive on digital content and digital services. In: Lohsse S, Schulze R, Staudenmayer D (eds) Data as counter-performance – Contract law 2.0?. Hart – Nomos, Oxford – Baden-Baden, pp 25–45

Resta G, Zeno-Zencovich V (2018) Volontà e consenso nella fruizione dei servizi in rete. Riv Trimest Dirit Proced Civ 72:411–440

Sattler A (2020) Autonomy or heteronomy – Proposal for a two-tier interpretation of Article 6 GDPR. In: Lohsse S, Schulze R, Staudenmayer D (eds) Data as counter-performance – contract law 2.0?. Hart – Nomos, Oxford – Baden-Baden, pp 225–251

Schmidt-Kessel M (2019) Consent for the processing of personal data and its relationship to contract. In: De Franceschi A, Schulze R (eds) Digital revolution – New challenges for law. Beck – Nomos, Oxford – Baden-Baden, pp 75–82

Schmidt-Kessel M (2020) Right to withdraw consent to data processing – the effect on the contract. In: Lohsse S, Schulze R, Staudenmayer D (eds) Data as counter-performance – Contract law 2.0?. Hart – Nomos, Oxford – Baden-Baden, pp 129–146

Schulze R (2016) Supply of digital content. A new challenge for European contract law. In: De Franceschi A (ed) European contract law and the digital single market – The implications of the digital revolution. Intersentia, Cambridge – Antwerp – Portland, pp 127–143

Sein K, Spindler G (2019) The new directive on contracts for the supply of digital content and digital services – scope of application and trader's obligation to supply – Part. 1. Eur Rev Contract Law 15(3):257–279

Sénéchal J (2020) Article 16(2) of the 'digital content and digital services' directive on the consequences of termination of contract, or the difficult articulation between union law on consumer contracts and union law on the protection of personal data. In: Lohsse S, Schulze R, Staudenmayer D (eds) Data as counter-performance – Contract law 2.0?. Hart – Nomos, Oxford – Baden-Baden, pp 147–162

Staudenmayer D (2020) Article 3 digital content directive (2019/770). In: Schulze R, Staudenmayer D (eds) EU Digital Law. Beck – Hart – Nomos, München – Oxford – Baden-Baden, pp 57–91

Twigg-Flesner C (2016) Disruptive technology – disrupted law? How the digital revolution affects (contract) law. In: De Franceschi A (ed), European contract law and the digital single market – The implications of the digital revolution. Intersentia, Cambridge – Antwerp – Portland, pp 21–48

Van Erp S (2020) Management as ownership of data. In: Lohsse S, Schulze R, Staudenmayer D (eds) Data as counter-performance – Contract law 2.0?. Hart – Nomos, Oxford – Baden-Baden, pp 77–93

Wendehorst C (2020) Personal data in data value chains – is data protection law fit for the data economy?. In: Lohsse S, Schulze R, Staudenmayer D (eds) Data as counter-performance – Contract law 2.0?. Hart – Nomos, Oxford – Baden-Baden, pp 193–223

Wendland M (2019) Sonderprivatrecht für digitale Güter. Zeitschrift für die Vergleichende Rechtswissenschaft 118:191–203

Zech H (2016) Data as a tradeable commodity. In: De Franceschi A (ed) European contract law and the digital single market – The implications of the digital revolution. Intersentia, Cambridge – Antwerp – Portland, pp 51–79

Zöchling-Jud B (2019) Das neue Europäische Gewährleistungsrecht für den Warenhandel. Zeitschrift für das Recht der Europäischen Union 16(3):115–133

Cookies and the Passive Role of the Data Subject

Andrea Maria Garofalo

1 Technological Features and Legal Regulation of Cookies

In recent years we have heard, more and more often, about cookies. But what exactly are they? And how are they legally regulated?

In a nutshell, cookies are small strings of text. During web browsing, each site sends cookies to the user's 'terminal equipment' (normally a computer or a smartphone). The browser stores the cookies so that it will be able to transmit them back if the user visits the same webpage again.

Cookies belong to several different typologies and can be classified in many different ways; however, there are still no standard categories in which to classify them, so each website adopts its own vocabulary and classification.

Nevertheless, the main distinction—which is between first-party and third-party cookies—is fairly widely known and obvious.

First-party cookies are those that are installed on the user's terminal equipment by the operator of the same website accessed by the user. These are the simplest cookies, which are sent by the operator of the site at the same time as the user's browser requests to load a specific webpage.

However, cookies can also be installed by other sites, such as in cases in which an image, a sound clip, or even just a link to a webpage or a pixel is embedded in a webpage by an entity other than the website provider. By loading such an image, sound clip, link or pixel, the browser sends a request to this entity, which also responds by installing a cookie on the user's terminal. These cookies are called 'third-party (cookies)'.

A. M. Garofalo (✉)
Ca' Foscari University of Venice, Venice, Italy
e-mail: andreamaria.garofalo@unive.it

R. Senigaglia et al. (eds.), *Privacy and Data Protection in Software Services*,
Services and Business Process Reengineering,
https://doi.org/10.1007/978-981-16-3049-1_7

Depending on the purpose for which they are installed, a further distinction is usually made between strictly necessary, functional (or functionality), analysis (or analytical), and profiling cookies for marketing purposes.

In reality, as mentioned, these classifications vary somewhat from operator to operator, just as it is uncertain whether each category belongs to technical or non-technical cookies. The latter represents another important distinction, corresponding to a different regulatory regime to which we shall later return.

Generally speaking, strictly necessary cookies are those which allow navigation on a site, for example, by avoiding the loss of settings already chosen or actions already carried out when one jumps from one page to another of the same website (which would, for example, require the choice of settings to be renewed each time or would render online shopping completely impossible).

Functional cookies are those that allow the site to remember the user and its preferences in order to personalise certain services requested by the user. For example, functional cookies allow a website provider to remember a user's choice of language or the location from which it connects (if the user wanted to share it, perhaps to be updated on the weather forecast). The 'shopping cart' on some sites may also work thanks to functional cookies, which allow a user to be remembered by a site even if they have not logged in with their account and even after they have closed it. Some sites' functionalities may be blocked if the user does not allow the use of these cookies (e.g., by browsing in private mode); in other cases, the user simply has to make the setting choices every time they enter the site (and not just every time they enter the site after closing the browser and, therefore, the whole session or even just once for a certain period).

Analysis cookies are those designed to allow the website operator to check the behaviour of users on its site. The analysis of the user's behaviour can take place in very different ways: at one end of the spectrum, there are cases where a website simply checks the time spent on each page and any loading problems on that page without in any way identifying the user (in which case cookies are not even necessary, and other technological tools can replace them); at the other end of the spectrum, there are cases where the website follows the user by monitoring their navigation and also checking how often and at what interval they access the site, as well as from where they access the site. In this case, it is necessary for cookies to associate a user ID with the individual user or, better, their browser, in order to recognise the user upon each new access and to follow them throughout their navigation on the website.

Finally, there are profiling (or targeting) cookies used for advertising purposes. These have various functions: they can measure the effectiveness of the advertisements shown in the banners of a certain website or remember that the user has visited that site even after leaving it (the user is thus followed during its navigation). In the latter case, the party that installed the cookies can know or at least assume what the user's tastes and interests are, especially if it also collects cookies while browsing other sites, and can thus choose the best advertising for him or her (the most likely to engage the user).

Each cookie may have a different expiration date pre-set in the browser depending on the privacy protocol it adopts. Consequently, cookies can also be categorised in terms of their duration, it being understood that the user can delete them whenever they want from their device through specific browser commands.

For example, functional cookies can be persistent or session cookies so that they last either forever (as a rule, however, for a maximum of two years) or only for the duration of a session (i.e., as long as the browser remains open).

From a legal point of view, cookies are primarily regulated by Directive 2002/58/EC ('e-Privacy Directive' or 'Cookie Law'), as amended by Directive 2006/24/EC and Directive 2009/136/EC, as well as—if they qualify as personal data—by Regulation EU 679/2016 ('GDPR').

According to the e-Privacy Directive, the use and installation of cookies are not permitted, except in the case of special categories of cookies or with the consent of the user. The rules of the Cookie Law apply irrespective of whether the cookies allow the identification of the user and therefore can be considered personal data. But which cookies exactly require consent under the e-Privacy Directive?

According to Article 5(3), "member States shall ensure that the storing of information, or the gaining of access to information already stored, in the terminal equipment of a subscriber or user is only allowed on condition that the subscriber or user concerned has given his or her consent, having been provided with clear and comprehensive information, in accordance with Directive 95/46/EC, inter alia, about the purposes of the processing. This shall not prevent any technical storage or access for the sole purpose of carrying out the transmission of a communication over an electronic communications network, or as strictly necessary in order for the provider of an information society service explicitly requested by the subscriber or user to provide the service".

Briefly, every cookie requires consent, unless it falls within the two mentioned exemptions (the 'communication' and 'strictly necessary' exemptions). Only under these circumstances, are cookies commonly called 'technical' and do not require any consent.

Strictly necessary cookies and functional cookies are mostly considered to be technical cookies since they are usually necessary for browsing the site or for services requested by the user. It is easy to understand the reason for this in the case of strictly necessary cookies; as for functional cookies, they usually concern services requested by the users themselves, such as setting the language of a site, indicating the place of connection in order to personalise the site, adding products to the list of favourites to remember the choice also in the future, and so on.

In these cases, the risk to the user's interests is absent or, in any case, it is certainly compensated by the benefit they receive. For example, without essential cookies, it would not even be possible to surf the internet; without functional cookies, the user would lose certain functionalities of the site that they have explicitly or implicitly shown they want to use.

Advertising profiling cookies, on the other hand, are not technical cookies, and indeed pose the greatest risks. Moreover, they are nowadays mostly third-party cookies, which are installed on the user's computer or smartphone when they visit a page on a website and which allow the third party (e.g., Google or Facebook) to follow the user throughout their navigation on that site and then to resell the advertising space of other sites or even their own advertising space to the operators of that site already visited by the user (who, therefore, has shown interest in the corresponding goods and services) or, in any case, to retailers of goods or services that may meet the user's desires (who has been profiled in this way).

The risk for the user, in this case, is very high since they are in fact spied on and followed in order to receive an advertisement that, precisely for this reason, will be very aggressive and will induce the user to purchase without having sufficiently thought it through. The risk is even higher when the advertising profiling is related not so much to the products and services bought as to the promotion of political ideas and opinions: in fact, the danger here is that a person becomes fixated on certain ideas, always receiving feedback from one side or faction, and makes choices without having really participated in the public debate. Scholars have coined the term 'surveillance capitalism' for this form of user control, which gives enormous power to the operators of websites and in particular social media platforms.

It is more complicated to qualify analysis cookies, especially if they are anonymous. They do not pose any particular risk to the rights and freedoms of users, so it would not seem necessary to make them subject to the user's consent. However, from a purely formal point of view, they are often third-party cookies (e.g., Google Analytics cookies), which qualify as personal (not anonymous) data.

National supervisory authorities themselves have a divergent position on this point.

Generally speaking, however, they all see the absurdity of fining those who use third-party analytics cookies without asking for consent, where the third party merely checks where and how the site is accessed. In this case, if the cookies are installed directly by the third party and if the latter does not use the users' data for other purposes, and makes them pseudonymous or anonymous when storing them (by changing part of the IP address digits), consent is not necessary or, in any case, its absence is not actually sanctioned. If, on the other hand, the analysis is carried out in conjunction with profiling and tracking techniques (including those designed to identify the profile of users of a website on a statistical basis), consent is again required. If then, the third party uses cookies for personal purposes, its role certainly becomes more intense and the risks for the rights and freedoms of the user are greater. In such a case, indeed, the third party itself becomes a joint controller.

The processing of cookies, as mentioned above, must also comply with the GDPR in cases where cookies can be considered personal data, i.e., where they allow the attribution of information to an identified or identifiable natural person. According to Article 4(1) GDPR, an identifiable person is a natural person who can be identified, directly or indirectly, by an online identifier, such as an IP address or a cookie associated with an IP address or user ID. For this reason, cookies for which consent is required by the e-Privacy Directive are virtually always considered personal data. Consequently, consent to cookies is generally deemed to be governed by the provisions on consent to the processing of personal data of which it constitutes a very particular application (as confirmed by Recital 17 and Article 2, (f), e-Privacy Directive and implicitly also by Recital 173 and Article 95 GDPR).

2 The Problem of Consent to Cookies

Cookies do not pose many problems from the point of view of verifying compliance with legal texts, which is easily verifiable even by a sufficiently experienced individual user. Control authorities can, therefore, without particular effort, check whether a website installs cookies without the necessary user consent when required.

The real deficiency in regulating cookies lies in the user's consent, which is affected—often in a particularly accentuated manner—by all the problems that, according to the literature developed on these issues, afflict consent to the processing of personal data. We should dwell on these for a moment.

It has been noted that the request for consent to the processing of personal data, and in particular that relating to cookies, usually involves a mere request to tick a box on a banner; and the act of consent usually consists simply of that simple tick. However, the exact explanation of the meaning of such consent is usually found in the information notice, which is normally very long and in any case too long to imagine that the data subject will read and analyse it before choosing whether or not to give consent.

Even if this were the case, the data subject would hardly understand it or, rather, they would hardly understand the risks that the use and installation of cookies bring with it. In fact, users usually underestimate the risks involved in the use of their data (risks linked not only and not so much to the curtailment of privacy but also to the limitation of other rights, such as the right to commercial self-determination), nor do they know the value of their personal data to their counterparts (so they are inclined to accept their disposal even without receiving any real benefit in return). What is more, users often assume that their data are already in circulation and therefore, since they can no longer exercise control over them, are indifferent to allowing new and further processing.

These phenomena can be gathered under the broad heading of 'privacy paradox': the more important data protection is in an evolving society, the less it is, on the one hand, structurally enforceable through the conducts of individuals and, on the other hand, perceived to be fundamental by those same individuals. We could conventionally distinguish the 'paradox of attention' from the 'paradox of evaluation': the former depends on the asymmetry of time between those who process data and those who provide them so that the users cannot be required to analytically read the conditions of the processing; the latter, on the other hand, on the lack of a widespread culture regarding personal data so that their processing is not—in short—felt as a threat to certain values.

These paradoxes, of course, make consent to the processing of personal data, and especially that relating to cookies, an ineffective regulatory tool, since often there is no real choice below consent. In the face of this, the main concern is to avoid cases in which consent is given without any real awareness on the part of the user.

In the face of these difficulties, the GDPR has intervened by regulating in general terms the consent to the processing of personal data with provisions that are also applicable to the case of consent to cookies and that are aimed precisely at overcoming the problems that have just been indicated.

According to Article 4(11), "'consent' of the data subject means any freely given, specific, informed and unambiguous indication of the data subject's wishes by which he or she, by a statement or by a clear affirmative action, signifies agreement to the processing of personal data relating to him or her". Article 7 states inter alia that "where processing is based on consent, the controller shall be able to demonstrate that the data subject has consented to processing of his or her personal data"; "if the data subject's consent is given in the context of a written declaration which also concerns other matters, the request for consent shall be presented in a manner which is clearly distinguishable from the other matters, in an intelligible and easily accessible form, using clear and plain language. Any part of such a declaration which constitutes an infringement of this Regulation shall not be binding"; "when assessing whether consent is freely given, utmost account shall be taken of whether, inter alia, the performance of a contract, including the provision of a service, is conditional on consent to the processing of personal data that is not necessary for the performance of that contract".

Let us see how these general provisions are applied to cookies. It is useful in this respect to refer to the case-law of the Court of Justice of the European Union (CJEU) and the Guidelines of the European Data Protection Board (EDPB).

First of all, the general principles and rules of the GDPR have been understood to mean that a website cannot make access to it conditional on consent to cookies. For example, cookie walls, i.e., screens that obscure the pages of sites and ask for cookie consent in order to gain access, are considered prohibited.

Moreover, since consent must be unambiguous, it is considered that simply scrolling down a page cannot be a sufficiently meaningful act. This also means that, in the absence of any determination by the user, cookies other than technical cookies cannot be installed on a user's computer or smartphone, not even while waiting for the user to decide to give or deny their consent.

Furthermore, it is considered that pre-checked consent boxes for cookies do not allow for valid consent, since they imply an opt-out mechanism, rather than an opt-in one (in this respect, see CJEU, Bundesverband der Verbraucherzentralen und Verbraucherverbände—Verbraucherzentrale Bundesverband e.V. v Planet49 GmbH, C-673/17).

Finally, it is considered that consent should be able to be given and refused for specific types of cookies, grouped according to their purpose: this would meet the requirement of granularity of consent. This, of course, does not prevent a situation of joint controllership from arising in respect of third-party cookies if the requirements are met (see CJEU, Fashion ID GmbH & Co.KG v Verbraucherzentrale NRW eV, C-40/17).

In view of this, a request for consent that appears in a banner listing the various cookies used by a certain website (session, functional, analysis, profiling) and requesting consent for those for which it is necessary (referring, for more details, to a more detailed information notice, with specific indication of the specific cookies and their duration period) is certainly compliant with the GDPR. The privacy notice may in fact be 'layered', which means at several levels of detail.

It is doubtful whether banners are allowed, which, in addition to the 'accept all cookies' button, do not contain a 'reject all cookies' button but only a link that redirects to pages where one can select the cookies they want and those they do not. On this point, the national Supervisory Authorities do not have a convergent position.

On the one hand, in fact, it might seem sufficient that the data subject is given the choice between all-inclusive consent or granular refusal (especially if, as necessary, navigation on the page does not coincide with implicit consent).

On the other hand, however, it may be argued that in order to make the banner disappear, the user may be led to click on 'accept all cookies', especially in the haste of browsing. This possibility leads to the view that the absence of a 'reject all' button or, even if present, a certain design that makes the 'accept all cookies' button more visible or in some way prominent, does not comply with the text of the GDPR.

3 Technological and Legal Future of Cookies

We have seen, very briefly, the regulatory tools adopted at the EU level to try to make effective consent for the processing of personal data and, in particular, to cookies.

However, we must admit that, despite these measures, the consent-based approach to cookies is still not entirely satisfying: in fact, consent as an approach remains ineffective.

Even today, faced with a request for consent to cookies, users typically end up accepting or rejecting all cookies, without making any specific choices. Users allow themselves to be led either by a tendency to be willing to provide their personal data or, on the contrary, by a concern for their privacy; and with respect to these two general preferences, only a judgement of the trustworthiness of the specific website plays a role. There is no time to really choose the cookies one wants, so one either accepts them all (often also without fully understanding the risks involved) or rejects them all (thus, however, making a choice that is undoubtedly disadvantageous for the website operator and that is not reasonable whenever the cookies are useful or at least not harmful for the user).

What is more, the choice often interrupts navigation in an annoying way, which, besides being a problem in itself, leads to another undesirable result: even where it is possible to either accept or reject all cookies, it is possible that, in the rush of browsing, the inexperienced or even the hurried user may be led to accept them. It can occur just because attention is first drawn to the 'accept all' button (even regardless of the design choices of the web page: simply, the expression 'accept' generally attracts more attention than the expression 'reject') or because 'accept all' appears to be the best choice in order not to incur a loss of website functionality or even not to interrupt navigation.

Due to the ineffectiveness of consent, individuals worried about their privacy more and more often decide to use forms of navigation (e.g., incognito or through

TOR) which avoid leaving footprints on the web. Moreover, browsers themselves are evolving towards automatic privacy protection, for instance, by detecting third-party cookies through specific algorithms and blocking them by default (this has already happened for Safari, thanks to the implementation of the new 2.3 version of WebKit's Intelligent Tracking Prevention and will soon happen for Chrome).

The idea behind this technological evolution is certainly appropriate: granular consent to cookies on each site is not the best form of regulation because, however well-regulated it may be, it remains an ineffective and inefficient protection tool. Consequently, it seems appropriate that choices on cookies should be made upstream once and for all, and that by default a choice corresponding to the refusal of consent should apply to all cookies that are unwanted.

However, the technological tools that are being developed to achieve these results have several unconvincing, if not even disadvantageous, aspects. On the one hand, there is the risk of indiscriminately closing the door to all sorts of cookies, even if they are useful not only for the site but also for the user, or in any case not harmful to the user (which could be the result of blocking all unnecessary cookies or at least all third-party cookies, without distinguishing them by category). On the other hand, the tendency to solve these problems through technological and not legal solutions lends itself to the risk that, on the programmers' side, new technological tools are developed to circumvent and overcome the limitations created. Let us look at these two aspects in more detail.

First, the limitation of third-party cookies achieved technologically disregards the fact that, for many sites, the use of these cookies is vital. The GDPR's current prohibition on websites making access conditional on the installation of cookies is hypocritical since it is based on the fact that consent is mostly given, although it can hardly be considered an actual act of will; anyway, this system can work, as long as most people do consent to the installation of cookies. Where, however, users deny consent on a widespread basis, e.g., through technological means that prevent the installation of all unnecessary or at least all third-party cookies, there could be a change in the whole architecture of the internet as it is known today, moving it from the forms of surveillance capitalism to other unknown forms. And this outcome might not be in line with the political objectives of the EU legal system, which tends to protect and supervise the market, rather than destroy it.

Hence the other possible danger: in order to maintain the current architecture of the internet, websites could adopt new technologies aimed at allowing user tracking again. In fact, such technologies are already in sight; they are, for example, based on fingerprinting or tracking technologies that are no longer on the client-side but server-side. These solutions, although more efficient and more secure, are necessarily less transparent, since they do not allow even the experienced user to know and control at any time—as happens with cookies—which entity is tracking him. Moreover, all this will further increase the overwhelming value of sites and platforms that do not need tracking technologies based on third-party cookies to monitor user behaviour, such as Amazon and Facebook, which know the preferences of their users irrespective of the use of cookies (thanks, for instance, to the control of shopping carts, purchases made, pages followed, or likes placed).

As we can see, the solutions that are spontaneously appearing on the horizon, outside of a regulatory control of the market, are not necessarily the best ones. But that is not all: even the proposal for a new e-Privacy Regulation, which was published in 2017 and in large part goes in the direction of these spontaneous solutions, is not entirely convincing. Although, the search for technological tools to overcome the difficulties present in the current situation (i.e., the ever-present request for consent to cookies) is certainly of merit.

In particular, Article 10(1), according to which "software placed on the market permitting electronic communications, including the retrieval and presentation of information on the internet, shall offer the option to prevent third parties from storing information on the terminal equipment of an end-user or processing information already stored on that equipment", is not entirely persuasive.

This solution largely reproduces the automatic blocking of third-party cookies that are already being adopted by the most recent updates of the most popular browsers (going even a little further, however, in prohibiting any processing operation regarding information already stored on that equipment). However, this provision is, on the one hand, not technologically neutral, and therefore lends itself to the risk of rapid obsolescence, and, on the other hand, also excessively rigid, as it does not make any attempt to balance the needs of users with those of websites.

The provisions of Articles 8(1) and 10(2) are more convincing.

Article 8(1) provides that consent is not required even for "the use of processing and storage capabilities of terminal equipment and the collection of information from end-users' terminal equipment (…) if it is necessary for web audience measuring, provided that such measurement is carried out by the provider of the information society service requested by the end-user". This overcomes the need for consent in a case where the use of cookies poses very little risk to the rights and interests of the user.

According to Article 10(2), "upon installation, the software shall inform the end-user about the privacy settings options and, to continue with the installation, require the end-user to consent to a setting". In this way, instead of indiscriminately prohibiting heterogeneous classes of cookies (such as third-party cookies), it is possible for the user themselves to decide which homogeneous classes of cookies they want to accept or reject by default. Such a consent, which is given once and upstream, is certainly more effective than a consent to be given each time a specific site is accessed.

This provision, however, needs to be completed in two ways: (a) it requires the creation of standard categories of cookies, which can be recognised as such by browsers (not only because they come from the website or third parties); and (b) it is necessary that the user is put in a position to be able to modify with a simple and quick choice the security settings for a specific website (in order to allow or prohibit, for instance, profiling).

The first need is felt particularly keenly today: not only are there independent institutions that are developing software to manage cookie consent, but even the Italian Supervisory Authority has stimulated public consultations aimed at creating a certain uniformity in the categorisation of cookies. The second need is in fact closely linked to the first: once standard categories are created, it will be possible to spread greater transparency and user awareness in the use of cookies; at the same time, it will be possible to foresee specific spaces in browsers aimed at detecting the request to install certain cookies, their automatic blocking or automatic acceptance, and, finally, the possibility of a different granular choice each time (i.e., for each site or each page of a site).

However, some doubts remain.

If this system becomes too demanding, website operators will, in any case, turn towards new forms of profiling, which may be even more dangerous than the current ones, primarily because they are less transparent (for instance, some companies offer profiling services based on complex probabilistic techniques).

If, on the other hand, the installation of cookies remains too widespread (compared to the risks involved and the effectiveness of the consent given by users), these regulatory choices will be considered to be at least partly ineffective.

In any case, it is likely that the future and ultimate solution will come about not through a reduced use of personal data but through a series of remedies outside data protection and belonging to other areas of law: for instance, transparency in the profile-based nature of certain advertisements or the allocation of a certain share of advertisements and news displayed to non-profile-based and random content, in order to recover, by means of the random provision of content, that freedom of the individual that would otherwise be compromised.

References

Agencia Española Protección Datos (2020) Guía sobre el uso de las cookies. https://www.aepd.es/sites/default/files/2020-07/guia-cookies.pdf

An Coimisiún Chosaint Sonraí-Data Protection Commission (2020a) Guidance note: cookies and other tracking technologies. https://www.dataprotection.ie/sites/default/files/uploads/2020-04/Guidance%20note%20on%20cookies%20and%20other%20tracking%20technologies.pdf

An Coimisiún Chosaint Sonraí-Data Protection Commission (2020b) Report by the data protection commission on the use of cookies and other tracking technologies. Following a sweep conducted between August 2019 and December 2019. https://www.dataprotection.ie/sites/default/files/uploads/2020-04/Report%20by%20the%20DPC%20on%20the%20use%20of%20cookies%20and%20other%20tracking%20technologies.pdf

Article 29 Data Protection Working Party (2013) Working document 02/2013 providing guidance on obtaining consent for cookies. https://ec.europa.eu/justice/article-29/documentation/opinion-recommendation/files/2013/wp208_en.pdf

Article 29 Data Protection Working Party (2012) Opinion 04/2012 on cookie consent exemption. https://ec.europa.eu/justice/article-29/documentation/opinion-recommendation/files/2012/wp194_en.pdf

Bond R (2012) The EU E-privacy directive and consent to cookies. Bus Lawyer 68:215–223

Bravo F (2017) Il consenso e le altre condizioni di liceità del trattamento di dati personali. In: Finocchiaro G (ed) Il nuovo Regolamento europeo sulla privacy e sulla protezione dei dati personali. Zanichelli, Bologna, pp 101–177

Buchner B, Kühling J (2018) Article 7 DS-GVO. In: Kühling J, Buchner B (eds) Datenschutz-Grundverordnung - Bundesdatenschutzgesetz - Kommentar, 2nd edn. C.H.Beck, München, pp 284–307

Caggia F (2019) Il consenso al trattamento dei dati personali nel diritto europeo. Rivista del diritto commerciale e del diritto generale delle obbligazioni 117(3):405–432

Commission nationale de l'informatique et des libertés (2020) Délibération n° 2020-092 du 17 septembre 2020 portant adoption d'une recommandation proposant des modalités pratiques de mise en conformité en cas de recours aux «cookies et autres traceurs». https://www.cnil.fr/sites/default/files/atoms/files/recommandation-cookies-et-autres-traceurs.pdf

Conseil Général de l'économie, de l'industrie, de l'énergie et des technologies (2018) Access to data, consent and the impact of the proposal for an ePrivacy regulation. https://www.economie.gouv.fr/files/files/directions_services/cge/e-privacy-EN.pdf

European Data Protection Board (2020) Guidelines 05/2020 on consent under Regulation 2016/679, version 1.1. https://edpb.europa.eu/sites/edpb/files/files/file1/edpb_guidelines_202005_consent_en.pdf

European Data Protection Board (2020) Guidelines 08/2020 on the targeting of social media users, version 1.0. https://edpb.europa.eu/our-work-tools/public-consultations-art-704/2020/guidelines-082020-targeting-social-media-users_it

European Data Protection Board (2019) Opinion 5/2019 on the interplay between the ePrivacy directive and the GDPR, in particular regarding the competence, tasks and powers of data protection authorities. https://edpb.europa.eu/sites/edpb/files/files/file1/201905_edpb_opinion_eprivacydir_gdpr_interplay_en_0.pdf

Garante per la protezione dei dati personali (2020) Linee guida sull'utilizzo di cookie e di altri strumenti di tracciamento. https://www.garanteprivacy.it/documents/10160/0/Consultazione+sulle+%E2%80%9CLinee+guida+sull%E2%80%99utilizzo+di+cookie+e+di+altri+strumenti+di+tracciamento%E2%80%9D+-+Allegato+1+-+Linee+guida.pdf/72eab081-e4c4-4500-77c3-8b6957f8cd12?version=2.0

Gatt L, Montanari R, Caggiano IA (2017) Consenso al trattamento dei dati personali e analisi giuridico-comportamentale. Spunti di riflessione sull'effettività della tutela dei dati personali. Politica del diritto 2:337–353

Heckmann D, Paschke A (2018) Artikel 7. In: Ehmann E, Selmayer M (eds) Datenschutz-Grundverordnung, 2nd edn. C.H. Beck-LexisNexis, München, pp 234–259

Ingold A (2018) Artikel 7. In: Sydow G (ed) Europäische Datenschutzgrundverordnung, 2nd edn. Nomos-Manz-Dike, Baden-Baden, pp 445–465

Klement JH (2019) Artikel 7. In: Simitis S, Hornung G, Spiecker I (eds) Datenschutzrecht. Nomos, Baden-Baden, pp 542–569

Konferenz der unabhängigen Datenschutzaufsichtsbehörden des Bundes und der Länder (2019) Orientierungshilfe der Aufsichtsbehörden für Anbieter von Telemedien. https://www.datenschutzkonferenz-online.de/media/oh/20190405_oh_tmg.pdf

Montanari M (2019) Article 7. In: Barba A, Pagliantini S (eds) Commentario del codice civile. Delle persone. Leggi collegate, vol. II. Utet, Torino, pp 134–143

Selinger E, Polonetsky J, Tene O (eds) (2018) The Cambridge handbook of consumer privacy. Cambridge University Press, Cambridge

Solove DJ (2013) Introduction: privacy self-management and the consent dilemma. Harv Law Rev 126:1880–1903

Thobani S (2020) Processing personal data and the role of consent. Eur J Privacy Law and Technol 93–104

Zuboff S (2019) The age of surveillance capitalism: the fight for a human future at the new frontier of power. Public Affairs, New York

Data Management Tools and Privacy by Design and by Default

Fabio Bravo

1 Data Management Tools (DMTs) and Main Legal Issues on Data Protection Law

The legal regulation in the matter of personal data protection has undergone relatively recent modifications through Reg. EU 679/2016 (GDPR), issued to replace Dir. 95/46/EEC. The new European regulation, applicable since 2018, must not be viewed as a finish point: we are in fact in the middle of an extensive regulatory evolution, bound to change radically in the near future, as can be clearly felt through the issuance of the Proposal for an EU Regulation in the matter of Data Governance of 25.11.2020 (Data Governance Act). This regulatory evolution has led to a significant paradigm shift: the long journey that led to the final establishment of a fundamental right to personal data protection (Article 8 Charter of Fundamental Rights of the EU) and its legal recognition within European legislation on data protection, then undertook a different path, in which matter the central issue becomes the usability, also of economic nature, of data and control over them. In this regard, the role played by the technological tools which manage data (known as Data Management Tools or DMTs) and allow not only the processing of personal data but also their management, sharing, control and usability, becomes paramount (Rundle 2006). They are therefore both data processing and data governance tools. With regard to the latter, DMTs can also be used to manage recognised rights in favour of the data subject, so as to achieve profitability from personal data. This phenomenon can be seen, in particular, in 'infomediation' relations (Bravo 2020, 2021; Hagel and Rayport 1997).

The GDPR has imposed the guarantee of the respect of the rights and freedoms of the data subject ever since the design of the personal data processing (privacy by design) and privacy by default, through the obligation to put in place specific

F. Bravo (✉)
University of Bologna, Bologna, Italy
e-mail: fabio.bravo@unibo.it

R. Senigaglia et al. (eds.), *Privacy and Data Protection in Software Services*,
Services and Business Process Reengineering,
https://doi.org/10.1007/978-981-16-3049-1_8

technical and organisational measures (Article 25 GDPR) (Bravo 2019; Bygrave 2017). This obligation, which also applies to DMTs, as to any (tool used for the) processing of personal data, entails significant interpretative problems, concerning both objective and subjective aspects.

DMTs can also be used to manage or enhance the data protection of the data subjects, they are therefore also at the centre of the debate on the enhancement of personal data protection and, in this regard, can take on Privacy Enhancing Technology (PET) functions. At the same time, DMTs can also be used as Privacy Management Tools (PMTs), with 'data governance' functions. They allow for both the exercise of rights of the data subject in the economic sphere (consent to processing in exchange for payment or another economic benefit; data portability; usability of data in the perspective indicated on the proposal for a regulation on data governance). Here the legal issues are different, and also concern the reification and capitalisation of data, their sharing also for commercial purposes and their use also for altruistic purposes.

2 Privacy by Design and by Default. Privacy Design Pattern and Privacy Dashboard

In Article 25(1) GDPR, it is said that the data controller shall "implement appropriate technical and organisational measures, such as pseudonymisation, which are designed to implement data-protection principles, such as data minimisation, in an effective manner and to integrate the necessary safeguards into the processing in order to meet the requirements of this Regulation and protect the rights of data subjects". These are measures that must be put in place not only "at the time of the processing itself" but also "at the time of the determination of the means for processing". These are therefore technical and organisational measures that the controller must undertake ever since the 'design' stage of the processing, taking into account a complex series of parameters set by the legislator, namely, "the state of the art, the cost of implementation and the nature, scope, context and purposes of processing as well as the risks of varying likelihood and severity for rights and freedoms of natural persons posed by the processing".

Prepared during the design stage, they must then accompany the processing throughout its entire life cycle. The controller is tasked with establishing what measures can be considered "appropriate" in accordance with the above-mentioned provision: among these, however, the legislator expressly includes, by way of example, 'pseudonymisation', meaning the set of processing operations appropriate to impede that personal data "be attributed to a specific data subject without the use of additional information, provided that such additional information is kept separately and is subject to technical and organisational measures to ensure that the personal data are not attributed to an identified or identifiable natural person" (Article 4(1)(5) GDPR). Thus, personal data are temporarily stripped of the references necessary to identify the data subjects and to make them identifiable, separating the additional information necessary for the identification. These must be kept separate and must undergo measures aimed at guaranteeing their non-traceability to a given subject (for example, through 'encryption'), if not through a reverse procedure (for example, a 'reversal of pseudonymisation').

During the design, the 'privacy design patterns' (or 'privacy patterns') are especially useful, which in the matter of privacy are a form of design patterns (Bravo 2019). Design patterns are generally used in the creation of software and can be understood as recurring code 'modules' or 'portions', to realise, with the appropriate readjustments where necessary, a certain function present in an application, without having to rewrite the code every time from scratch. Among the communities of developers, 'pattern libraries' emerge, which can be consulted to more rapidly enter a given function, without having to rewrite it every time from scratch. The use of such design patterns, with various contents, therefore affects the architecture itself of the software, which can benefit from ready-made and already tested 'modules' for the given use one seeks, envisaged during the design stage of the software. These design modalities may comply with the concept of privacy by design provided for in Article 25(1) GDPR (Bravo 2019).

The new Regulation aims at resorting to a protection solution of personal data protection which can be reached during the design stage, so as to put in place an architecture (of the software, but also of the device, of the computer system as a whole, etc.) capable of offering adequate guarantees for the protection of personal data (e.g., privacypatterns.org, privacypatterns.eu). Thus, one of the most useful applications, in this direction, comes from the use of privacy design patterns, meaning the modular portions of 'code' with which programmers or developers create—and make available to the community of programmers and developers, for subsequent reuse—certain functions aimed at structurally ensuring the compliance (Reidenberg 1998) with personal data protection. In this regard online there are also portals—managed in the matter of international research projects carried out at the university level—aiming to conduct censuses on and disseminate privacy patterns, in order to spread their use and improve their functions.

The concept behind these projects is to develop and promote the use of technologies capable of ensuring compliance with personal data protection, identifying also the correct methodology to adopt to create privacy patterns, in reference to which one must keep into account not only the more IT-related aspects concerning the 'code' creation but also legal aspects, given the difficulty in translating into computer language the regulatory data and the (legal) *principles* regulating the matter under examination.

The procedure leading to the provision of technological tools, in the prospect evoked by Article 25 GDPR, is made up of different stages: the design does not in fact constitute the initial stage, as it is preceded by the *analysis* stage, aimed at determining the characteristics of the system and the aspects concerning personal data protection, based on two possible approaches (*risk-based approach* or *goal-oriented approach*). Resorting to predetermined privacy patterns, and also the development of new privacy patterns, deploys all its efficacy thanks to the replicability of adaptable 'methodologies' and 'models', during the design stage, to the different systems used for personal data processing. Privacy patterns fall under the realisation modalities of what has been defined as 'privacy by design' (or 'data protection by design') (Cavoukian 2009; Pagallo 2012), belonging to the more complex type of technologies with which one seeks to realise or enhance the protection requirements of personal data in the computer system used for processing (known as Privacy Enhancing Technologies, PETs) (Burkert 1997).

Along with data protection by design, the European legislator also regulated data protection by default, laying down that the data controller shall "implement appropriate technical and organisational measures for ensuring that, by default, only personal data which are necessary for each specific purpose of the processing are processed" (Article 25(2) GDPR).

This protection technique is complementary to the first, based on the use of procedures and technologies protecting one's privacy, resorting to data-oriented strategies: no intervention is required on the structure of the software, the device or the computer system or the processing, in its entirety; an intervention entailing a 'control' on personal data processing is instead required, both during the acquisition stage and the subsequent reprocessing, employing technical and organisational measures (Crespo Garcia et al. 2015). In particular, the application of the regulation under examination requires that the technologies devised during the design stage for privacy protection—therefore present in the computer system in compliance with the obligation referred to in Article 25(1) GDPR—be then 'pre-set' by default to limit processing solely to the *necessary data* for achieving the purposes of the processing.

Data protection by default, outlined in Article 25(2) GDPR, focuses its attention solely on the setup of the control over the data subject to processing, requiring that they be 'filtered' by default for the entire life cycle of the processing, ever since their acquisition, through technical and organisational-procedural measures. The default setup (*ex-ante*) does not exclude a different setup in a subsequent stage, as is clarified in the previous examples: the default setup ordered by the European legislator seems to be oriented toward a 'dynamic' use of Privacy Enhancing Technologies (PETs), in that the level of protection ensured by them varies depending on "each specific purpose of the processing" the controller decides to legitimately realise. To assess what are the sole *necessary data* that the processing, as default setup, must consider, Article 25(2) GDPR sets qualitative and quantitative criteria, always to be estimated in relation to the "purposes", and expressly provides for "that obligation applies to the amount of personal data collected, the extent of their processing, the period of their storage and their accessibility". It would therefore be useful to allow the data controller and the data subject to benefit from a privacy dashboard to set up the 'privacy options' ensuring, by default, the greatest protection for the data subject, which can then be modified in relation to the requirements and purposes pursued, allowing a subsequent intervention on the privacy settings of the Privacy Management Tools, where necessary.

3 Application Issues

The implementation obligation of privacy by design and by default (Article 25 GDPR) must be carried out by the data controller by implementing the accountability principle (Articles 5(2), 24(1) GDPR), whereby the data controller shall guarantee and be able to prove, through adequate and, where necessary, updated technical and organisational measures that the processing is performed in compliance with the provisions of the GDPR. Navigating this is not easy in light of remarkable application issues entailed in the formulation of Article 25 GDPR.

3.1 Excessive Vagueness of the Obligation and Difficult Identification of Contents

The first problem concerns the excessive vagueness of the regulation in question, especially if compared with the different solution in the "Proposal for a Regulation of the European Parliament and of the Council on the protection of individuals with regard to the processing of personal data and on the free movement of such data" (25.01.2012, COM (2012) 11 final).

Indeed, in Article 23 of this proposal—corresponding to current Article 25 GDPR—there were two paragraphs eliminated in the final text, in which, after the requirements aimed at guaranteeing the privacy by design and by default obligations (Articles 23(1) and 23(2) Proposal) the Commission being "empowered to adopt delegated acts (…) for the purpose of specifying any further criteria and requirements for data protection by design requirements applicable across sectors, products and services" (Article 23(3) Proposal) was contemplated, together with the further power to "lay down technical standards for the requirements laid down in paragraph 1 and 2" (Article 23(4) Proposal).

In the final text of the regulation in question, the powers of the Commission were drastically diminished, thus now the GDPR merely includes (in Article 25 GDPR) the obligation to adopt technical and organisational measures aimed at protection starting from the design and by default setting, delegating said measures directly to the subject which sets up the technological structure necessary for the personal data processing.

The step back by the European legislator has been justified owing to the difficulty in legitimising, pursuant to Article 290(1) TFEU, the powers delegated to the Commission, admitted solely for "non-legislative" acts of general scope, which integrate or modify certain "non-essential" elements of a legislative act, explicitly defining objectives, content, duration, scope and conditions of the delegation (Koops and Leenes 2013). Therefore current Article 25 GDPR, which has remained vague in its content and impossible to integrate at an institutional level, was deemed overly generic, which makes the precept non-implementable in all its possible applications, if not even evanescent: the translation of the regulation into a computer 'code' (and, before that, into an algorithm) becomes extremely complicated, if not concretely non-implementable, as there are multiple and diverse modalities of interpreting and concretely implementing the provisions of current Article 25 GDPR.

3.2 Doubts on the Existence of the 'Hardcoding' Obligation

There is another issue related to Article 25 GDPR. The formulation of the regulation has been perhaps overestimated owing to the emphasis of the 'privacy by design' principle in the debate concerning *Privacy Enhancing Technologies* (PETs): in fact, it should be scaled down, in that performing 'architectural' interventions on the software or the hardware used for personal data processing in order to guarantee compliance with the regulation in the matter of data protection (known as hard-coding) is not an operation that can be concretely translated into operative practice, given the various obstacles undermining its feasibility (therefore, it has been maintained that "Privacy regulation cannot be hardcoded", Koops and Leenes 2013). This argumentation is largely acceptable.

One difficulty is due to the complexity of the regulatory system concerning personal data protection, which cannot be limited solely to the GDPR provisions: at a European level, for example, Article 8 of the Charter of Fundamental Rights of the EU, Directive 2002/58/EC (known as e-Privacy Directive) subject to review (it will be replaced by a Regulation, complementary to the GDPR). For example, the next issuance of the EU Regulation on "Data Governance" (still as a proposal). At the national level, instead, there are other provisions that may become relevant in the matter of data protection, albeit not contained in said Regulation (see, for example, the provisions concerning remote control of workers).

Other significant matters include the technical impossibility of translating the 'legal' rule into a 'technical' rule integrated with the software or hardware, owing to the modalities in which it is formulated, the need for interpretation, often not univocal and changeable over time, also in relation with the court decisions or the decisions by the supervisory authority (Koops and Leenes 2013).

The critical aspects do not, however, lead to complete dismantling, in that hard-coding can be performed on computer systems tasked with personal data processing due to certain well-defined and easily identifiable legal rules (such as the use of encryption, data pseudonymisation, specific technologies for access control to a computer system and the fields of the allowed processing, etc.), but not as a general mandatory rule (Koops and Leenes 2013).

Moreover, the stress of the legal theory and the institutions on technological constraints—suggested by the use of the term Privacy Enhancing Technologies (PETs)—seemed excessive, when considering that the 'protective obligations' outlined by Article 25 GDPR do not solely entail "technical" but also "organisational" measures, meaning that the principles of privacy by design and by default can be achieved also by resorting to a less rigid interpretation of the Regulation in question (Koops and Leenes 2013).

3.3 On the Appropriateness Criteria of the Measures

As already specified, Article 25(1) GDPR begins by establishing that the obligation of the data controller to put in place "*appropriate*" technical and organisational measures—aimed at effectively implementing the principles of data protection and integrating into the processing the necessary guarantees to both meet the requirements referred to in the Regulation and protect the rights of the data subjects—must be complied with "taking into account the state of the art, the cost of implementation and the nature, scope, context and purposes of processing as well as the risks of varying likelihood and severity for rights and freedoms of natural persons posed by the processing (…)".

These are not *selective* criteria with regard to the application or lack thereof of the obligation of the measures in question, which must in any case be met, but rather *assessment* criteria concerning the appropriateness (and congruity) of the measures to be applied, which—if left to decide solely to the controller—may be overly evanescent, if not defined within a self-disciplinary regulatory framework (codes of conduct and certification mechanism) and, above all, by resorting to "standard models and (…) operational schemes put in place at the initiative (and under the leadership) of the European Data Protection Board (Article 70), whose contribution promises to be decisive so that, in the tension between people's rights and technological evolution, it is technologies and economic practices that conform to the institutions and concepts of personal data protection, and not these, and the underlying concepts, to merely have to adapt to the former to allow the development of certain technological applications" (D'Orazio 2016).

Thus, the critical issues related to the risks of not determining these criteria, detectable in any case during the initial stage, are bound to be gradually overcome in an objective manner, although one can at any rate resort to the 'reasonableness' principle to hermeneutically guide the interpreter in the concrete application of the above-mentioned assessment elements *ex* Article 25 GDPR.

3.4 On the Recipients of the Obligation

Another complex issue concerns the correct identification of the recipients of the obligation referred to in Article 25 GDPR. The problem had already been highlighted, in Italy, with regard to the application of Article 3 Italian Legislative Decree 196/2003 (Italian Privacy Code), which however was characterised by an 'impersonal' formulation, i.e., without specifying the subject to whom the provision was to be considered applicable ("The information systems and the computer programmes are configured…"). Although it was clear that the first recipient of this regulation was the processing controller, solutions have been put forward to extend the applicative scope to include also producers of hardware and software used in the processing (Buttarelli 2007), which can also be traced in the measures of the Italian Supervisory

Authority (e.g., Italian Data Protection Authority, provision on the measures to adopt for the legitimate use of videophones, 20.01.2005, doc. web n. 1089812, para. 4). It is however an extension of the applicative scope of the regulation which appears to act more with regard to moral suasion than to the compliance with a mandatory provision, given that the manufacturers—and the providers—of technological tools used by the controller to process other persons' personal data are not—normally—controllers themselves of the processed data (big players of IT—such as Google and Facebook—do not fall under this category, as they process personal data through technological tools they themselves designed, manufactured and used: in this case, the producer of the technology can also be the data controller, but this evaluation must, at any rate, be made in each single concrete case).

With regard to the obligations to adopt the technical and organisational measures referred to in Article 25 GDPR, in particular in pursuance of the privacy by design principle, similar issues arise, which the Article 29 Working Party has already sought to solve with "Opinion 02/2013 on apps on smart devices" of 27.02.2013, highlighting the cases in which manufacturers of operating systems (OSs) and devices can be considered 'data controllers' or 'joint controllers'): "The OS and device manufacturers should also be considered as data controllers (and where relevant, as *joint controllers*) for any personal data which is processed for their own purposes such as the smooth running of the device, security, etc. This would include user-generated data (e.g., user details at registration), data automatically generated by the device (e.g., if the device has a 'phone home' functionality for its whereabouts) or personal data processed by the OS or device manufacturer resulting from the installation or use of apps. Where the OS or device manufacturer provides additional functionality such as a backup or remote locate facility they would also be the *data controller* for personal data processed for this purpose. Apps that require access to geolocation must use the location services of the OS. When an app uses geolocation, the OS may collect personal data to provide the geolocation information to the apps and may also consider using the data to improve its own location services. For this latter purpose, the OS is the *data controller*".

As can also be noted by discussing the above-mentioned specifications of the WP29, the application of the obligations referred to in Article 25 GDPR in the matter of privacy by design to the manufacturers of technological tools is neither generalised nor automatic: it may occur that a data controller may decide he or she must use a device or a software of 'third parties' to process personal data, as they remain unrelated to the processing in question, they cannot be considered directly recipients of said provision (which, unlike Article 3 of Italian Privacy Code is not built in an impersonal manner, but refers its precept, textually, to the 'controller').

In other words, the GDPR cannot be applied to manufacturers of technological tools unless they entail, at least partially, the quality of controllers or joint-controllers of the processing, therefore, where this condition—which must be ascertained case by case—is not met, data protection by design seems to be bound to remain a sort of empty box, given the inapplicability of the obligations to the subjects who, during the design stage, can affect the compliance of the technological tool with the regulatory provisions. This seems to be confirmed also by the content of Recital 78 GDPR, where—after tracing back the data protection by design and by default obligation to the field of the more general security obligation referred to in Article 32 GDPR—is addressed to the 'manufacturers' not as direct recipients of these obligations, but as subjects toward whom an action of 'encouragement' must be made, based on the logic of moral suasion, which is not unrelated to 'social responsibility' (CSR).

Although in these cases the regulation in Article 25 GDPR is not directly applicable to the manufacturers, it can be understood as to impose to the data controllers, to resort (solely) to technological tools which are compliant with the regulation in the matter of personal data protection, for which manufacturers, during the design stage, have adjusted to the rationale referred to in the provision in question. In discussing Article 25(1) GDPR, it must not be forgotten that "at the time of the determination of the means for processing", "the controller" shall "implement appropriate technical and organisational measures (…) which are designed to implement data-protection principles (…), "taking into account the state of the art". Thus, the data controller who cannot technically intervene during the design determining the 'privacy-compliant' architecture of the tool to use would nevertheless be obliged, while determining the means to use for the processing, to organise oneself to effectively implement the privacy (or data protection) by design principle, resorting to tools which, based on the state of the art, have been prepared by the manufacturer in compliance with the requirements of the Regulation.

Therefore, if the obligation to put in place the 'privacy-compliant' tool to use for processing personal data is not immediately applicable to the manufacturers, the same result can be reached if one considers that the controller (different from the manufacturer) is at any rate required, while determining the means to use for the processing, to select the tools with the above-mentioned compliance characteristics.

4 A Look Forward: European Data Governance

With the Proposal for a Regulation on Data Governance, the EU embraces the prospect of the control of personal data for purposes that are commercial, of public interest or 'altruistic'. The reuse in the EU of certain data categories held by public entities, voluntary registration systems for entities that collect and process data made available for altruistic purposes, and "a notification and supervisory framework for the provision of data sharing services" (Article 1), meaning "the provision by a data holder of data to a data user for the purpose of joint or individual use of the shared data, based on voluntary agreements, directly or through an intermediary" (Article 2(1)(7)) are regulated.

The notification must be made by the service provider to the competent national authority, which shall submit it to the authorities of the other Member States and to the European Commission, which will keep a register of data sharing service providers. The notification shall be a necessary requirement for the performance of the sharing service (Bravo 2021), which shall then be provided only if further conditions are met, including the ban from using the data "for other purposes than to put them at the disposal of data users"; the obligation to use "the metadata collected from the provision of the data-sharing service (…) only for the development of that service"; respecting the competitive dynamics; pursuing the best interest of the subjects receiving the service; the obligation of guaranteeing high-level security, of guaranteeing continuity in the service provision and access to the data on the part of the "data holders" and "data users" (Article 11).

The national authorities shall conduct the supervision of the provision of the data-sharing service (Bravo 2021) and, in the event of violations of the regulation in question, they can impose "dissuasive financial penalties which may include periodic penalties with retroactive effect" and put in place the "cessation or postponement of the provision of the data-sharing service".

The response of the European legal system to the datafication process of society (and of the economy) is underpinned by the enhancement of the supervision regarding data protection, where data traffic, also thanks to DMTs, is increasingly present in market dynamics.

References

Bravo F (2019) L'«architettura» del trattamento e la sicurezza dei dati e dei sistemi. In: Cuffaro V, D'Orazio R, Ricciuto V (eds) I dati personali nel diritto europeo. Giappichelli, Torino, pp 775–854

Bravo F (2020) Il commercio elettronico dei dati personali. In: Pasquino T, Rizzo A, Tescaro M (eds) Questioni attuali in tema di commercio elettronico. Edizioni Scientifiche Italiane, Napoli, pp 83–130

Bravo F (2021) Intermediazione di dati personali e servizi di data sharing dal GDPR al Data Governance Act. Contratto e impresa Europa 1(1):199-256

Burkert H (1997) Privacy-enhancing technologies. Typology, critique, vision. In: Agre PE, Rotenberg M (eds) Technology and privacy. The new landscape. San Diego, California, US, pp 125–142

Buttarelli G (2007) Commento sub art. 3 d.lgs. 196/20013. In: Bianca CM, Busnelli FD (eds) La protezione dei dati personali. Commentario al D.Lgs. 30 giugno 2003, n. 196 («Codice della privacy»). Cedam, Padova, I, pp 32–40

Bygrave LA (2017) Data protection by design and by default: deciphering the EU's legislative requirements. Oslo Law Rev 4(2):105–120

Cavoukian A (2009) Privacy by design. The 7 foundational principles, Ottawa. http://www.ipc.on.ca/wp-content/uploads/Resources/7foundationalprinciples.pdf. Accessed 13 January 2021

Crespo Garcia A et al (2015) Privacy- and security-by-design methodology handbook, vol 1.0, 31 Dec 2015. http://www.trialog.com/wp-content/uploads/2018/02/PRIPARE-Methodology-Handbook-Final-Feb-24-2016.pdf

D'Orazio R (2016) Protezione dei dati by default e by design. In: Sica S, D'Antonio V, Riccio GM (eds) La nuova disciplina europea della privacy. Wolters Kluwer-Cedam, Milano, pp 79–110

Hagel J, Rayport JF (1997) The new infomediaries. Mckinsey Quart 4:54–70

Koops BJ, Leenes RE (2013) Privacy regulation cannot be hardcoded: a critical comment on the "privacy by design" provision in data-protection law. Int Rev Law Comput Technol 28(2):159–171

Pagallo U (2012) On the principle of privacy by design and its limits: technology, ethics and the rule of law. In: Gutwirth S, Leenes R, de Hert P, Poullet Y (eds) European data protection: in good health? Springer Science & Business Media, pp 331–346

Reidenberg JR (1998) Lex informatica: the formulation of information policy rules through technology. Texas Law Rev 76(3):553–593

Rundle MC (2006) International personal data protection and digital identity management tools. Berkman Center Research Publication No. 2006-06, Available via SSRN. https://ssrn.com/abstract=911607. Accessed 13 January 2021

Reconciling Data Protection and Cybersecurity: An Operational Approach for Business Sector

Alessandro Mantelero and Giuseppe Vaciago

1 The Legal Framework. A Business Perspective

There are several sector-specific analyses of EU data protection and cybersecurity directives and regulations, but the different legal fields have prevented the development of a joint analysis of these different sets of provisions from a business perspective.

This artificial distinction between data protection and cybersecurity has led the legal analysis to consider as separate a number of obligations and procedures that are often deeply connected in the daily business activities of many companies. Moreover, a sector-specific standpoint fails to reveal the common approach of EU regulation and acts as an obstacle to the development of an integrated model for legal compliance.

Although data protection provisions focus on personal data and data subject's rights, the main objective of the legislator is to prevent potential risks related to the use of personal data. In this regard, Regulation 2016/679 (GDPR) provides a general framework, defining and stating the main binding principles for data use and data security (such as data minimisation, storage limitation and data confidentiality) that shape the entire edifice.

The GDPR takes a principles-based approach that is crucial in establishing a basic paradigm, but this paradigm needs to be further elaborated through examination of the other regulations, with a more technology-based and context-specific focus. From this

All authors have contributed equally. The authors thank Dr. Maria Samantha Esposito and Ms. Nicole Monte for their outstanding research assistance.

A. Mantelero (✉)
Polytechnic University of Turin, Turin, Italy
e-mail: alessandro.mantelero@polito.it

G. Vaciago
University of Insubria, Varese, Italy
e-mail: giuseppe.vaciago@uninsubria.it

R. Senigaglia et al. (eds.), *Privacy and Data Protection in Software Services*,
Services and Business Process Reengineering,
https://doi.org/10.1007/978-981-16-3049-1_9

perspective, the general framework provided by this Regulation must be coordinated with the various sector-specific regulations, which apply these regulatory principles in detail.

In our analysis, we therefore include three other legal instruments that regulate crucial areas of data processing in business activities (i.e., payments, identification and network and information systems security): Directive (EU) 2015/2366 (PSD2), Regulation (EU) 910/2014 (eIDAS), and Directive (EU) 2016/1148 (NIS).

PSD2 Directive on payment services provides for the introduction of Third-Party Providers (PISPs and AISPs) as new payment services with permission to access users' accounts. PSD2 therefore requires technologies that can ensure the protection of financial data, safeguarding both personal data and the security of the whole payment services environment.

Similarly, eIDAS Regulation on electronic identification and trust services for electronic transactions aims to create a secure environment for data subjects, their information, and rights, with regard to digital identity and access to personal data through digital identity, facilitating electronic interactions between businesses, citizens and public authorities, and providing a legal framework for the exchange of identity (electronic signatures and seals).

The same approach centred on technical and procedural rules is common to the NIS Directive on security of network and information systems, which focuses on international coordination and creates new institutional cybersecurity bodies with a broader scope than data protection (Markopoulou-Papakonstantinou-de Hert 2019). This directive contributes significantly to creating a safe environment for data processing and information sharing, in particular with regard to the essential services sector and providers of network and information systems.

In considering these different legal instruments and their role in boosting data protection and cybersecurity, the following sections do not adopt a regulatory-focused standpoint—centred on each of the different regulatory instruments considered—, but a cross-cutting view focusing on the main building blocks of an operational approach to these instruments, combining data protection and cybersecurity.

To this end, we will highlight the relationships between the legal provisions and the associated technological and organisational measures, with the aim to identify the key operational, legal and technical, elements of the European approach to data regulation in the business sector.

Four main thematic areas are investigated: dataset composition; risk assessment and security management; reporting obligations and mitigation measures; business continuity, disaster recovery, and resilience. This thematic order reflected in the following sections is not arbitrary but reproduces the different steps that should be put in place by each data controller in conducting a data-centred activity.

From the initial data collection to the ex-post verifications on legal compliance, the following sections point out the interplay between the various domain-specific provisions and their concrete implementation. A systematic overview of this interaction is provided in the concluding remarks, highlighting the main regulatory drives that underpin the EU framework in the field of data protection and cybersecurity.

2 Datasets: Data Minimisation, Storage Limitation, and Confidentiality

Datasets play a central role in digital services, increased by the massive process of datafication that has characterised the last decade. Data quality is therefore crucial in terms of product design and process, product quality and minimising potential adverse impact on data subjects. From an operational perspective, the following three principles set out in the GDPR should guide data management, regardless of the specific context of data use: data minimisation, data storage limitation, and data confidentiality.

Regarding data minimisation and data storage limitation, the GDPR requires controllers to limit the amount of data processed to the strictly necessary and do not retain personal data for longer than required by the purposes for which they were collected or further processed. Controllers should therefore define the relevant data retention period and adopt systems to automatically delete the data after this period has expired.

From a cybersecurity perspective, a strategy focused on data minimisation and storage limitation can help reduce the impact of data breaches resulting from cyberattacks or incidents (Mantelero–Vaciago 2017).

Regarding data confidentiality, the GDPR contains a set of provisions concerning access control and security, based on a more general approach adopted by data protection law over the years which emphasises the role of task distribution through the definition of roles and responsibilities of the entities involved. Moreover, both controllers and data processors must put in place specific measures to ensure a level of security appropriate to the risk (Article 32 GDPR; ENISA 2017a, b).

These organisational measures should be implemented together with the technical ones. Firms should adopt applications that allow them to create, approve, review and delete user accounts (ENISA 2016). The use of log files is also an essential security measure, enabling the identification and tracking of user actions, and helping to identify potential internal and external attempts at system violation.

To prevent data loss, destruction, or damage, it is important to ensure server and database security, as well as network and communication security. Several measures can be taken in this respect (e.g., anti-virus, malware detection systems), including monitoring of traffic to and from IT systems (e.g., firewalls and Intrusion Detection Systems).

The physical security of systems should also be taken into account to ensure a secure operating environment (e.g., ID Badges for personnel and visitors accessing the premises of the organisation, physical barriers, automatic fire suppression systems, continuous power supply, etc.).

Finally, hiding personal data and their interrelationships from plain view may also be useful to prevent data being acquired and misused by unauthorised actors. In this regard, the GDPR explicitly refers to pseudonymisation and encryption (ENISA 2021).

3 Risk Assessment and Security Management

While dataset creation and their composition play an important role in data-driven solutions, a core area where the interplay between data protection and cybersecurity is more evident concerns risk assessment. Here, the rights-oriented paradigm of the GDPR and the security-centric approach of cybersecurity rules are combined in providing a set of provisions that prevent adverse effects on systems and individuals in the context of today's socio-technical systems, where these components are inevitably intertwined.

The GDPR adopts a scalable approach to risk assessment, from a less structured assessment to a broad in-depth analysis (Articles 35 and 36), with several obligations that have a procedural impact on the use of data, from the initial assessment phase to the concrete implementation of the outcome of the assessment.

The GDPR does not stipulate a specific set of security measures but rather requires data controllers and, where applicable, data processors, to adopt appropriate technical and organisational measures to protect personal data. However, the Regulation provides some recommendations as to what type of security measures may be considered 'appropriate', referring to specific technologies (e.g., pseudonymisation and encryption) and to procedural approaches, including regularly testing and assessing the effectiveness of technical and organisational measures adopted (Article 32).

A broader perspective is then adopted in the provisions concerning the Data Protection Impact Assessment (DPIA), which goes beyond data security and takes a more holistic risk-based approach focusing on the impact of data use on the rights and freedoms of natural persons (Raab 2020; Gellert 2020; Mantelero 2018; Article 29 Working Party 2017).

Based on the risk assessment, data controllers should put into place technical and organisational measures to implement data protection in an effective manner, to integrate the necessary safeguards with the processing and set any pre-existing configuration value or processing option in line with the principles of data minimisation and purpose limitation (data protection by design and by default: EDPB 2019; Jasmontaite et al. 2018; Bygrave 2017; Le Métayer 2010; EDPS 2018).

Finally, regarding assessment of security measures (Article 32.1.d), businesses should carry out vulnerability assessments, as well as application and infrastructure penetration tests, but the GDPR does not specify any particular techniques for this purpose (e.g., software to test connections to outside networks and look for gaps in configuration and ethical hacking).

Risk and security management requirements are more specific in the Payment Services Directive (PSD2): payment institutions are required to provide specific information on the procedure in place to monitor, handle and follow-up any security incident or security-related customer complaint.

This is coherent with the need to implement specific cybersecurity technical standards which requires, on the one hand, risk management, security readiness and incident response preparedness in reducing the risks and consequences of major cyber and physical events, including (1) an adequate corporate governance structure; (2) security policies and incident response plans, procedures and toolkits; (3) information sharing arrangements with government agencies and industry centres; (4) table-top exercises; (5) third-party vendor contracts and management; (6) insider threat programmes; and (7) employee training programmes.

Another context-specific implementation of risk and security management principles relating to personal information is put in place by the eIDAS Regulation, which requires the adoption of electronic identification schemes, authentication mechanisms and a supervisory authority. All these elements on the one hand create a risk-adverse architecture and, on the other hand, facilitate the adoption of procedural solutions to reduce potential negative consequences with regard to digital identity and access to personal data through digital identity.

eIDAS Regulation (Article 19) refers to the obligation to adopt 'appropriate' technical and organisational measures to manage the risks associated with the security of the trust services provided. This may concretely mean, for example:

(i) adopting all measures to ensure an appropriate and adequate level of security of all trust services and products used and provided, which process personal data, particularly in the context of electronic identification schemes requiring a high degree of interoperability between the different systems adopted in EU states (i.e., authentication factors knowledge-based, possession-based, or biometric-based);
(ii) Implement a single IT incident management process in order to notify the supervisory body within 24 hours and the supervisory authority within 72 hours of becoming aware of the incident, as well as informing those involved (data subjects, service users) in the cases provided for.

Risk management and data security principles, generally set out in the GDPR and specified in the two contexts examined, receive more emphasis in the NIS Directive because of its four principal goals: manage security risk; protect against cyberattack; detect cybersecurity events; and minimise the impact of cybersecurity incidents. Not surprisingly, this Directive also provides a notion of risk, which is defined as "any reasonably identifiable circumstance or event having a potential adverse effect on the security of network and information systems".

With a view to raising the common security level of network and information systems across the EU, the NIS Directive (i) lays down obligations on all Member States to adopt a national strategy on network and information systems security; (ii) creates a Cooperation Group to support and facilitate strategic cooperation and the exchange of information among Member States and foster trust and confidence amongst them; (iii) creates a network of Computer Security Incident Response Teams (CSIRTs network) to further contribute to the growth of trust and confidence; (iv) establish security and notification requirements for operators of essential services and digital service providers; and (v) lays down obligations for Member States to designate national competent authorities, single points of contact and CSIRTs.

From an organisational perspective, the Directive provides a framework for strategic objectives and priorities on the security of network and information systems at a national level. It lists the significant elements on which Member States' national strategies should focus with regard to risk and security management, defining specific objectives and priorities to be achieved through a governance framework with assigned roles and responsibilities to government bodies and other actors. These strategies must adopt an assessment plan to identify potential risks, including the measures relating to preparedness, response and recovery.

In terms of concrete implementation of these strategies, the Directive specifies technical and organisational measures to manage security risks to network and information systems (Article 14). Operators of essential services must put in place appropriate technical and organisational measures to prevent and minimise the impact of incidents affecting the security of the systems.

Article 23 of the Directive requires the European Commission to review the functioning of this Directive periodically. As a result of the review process, a new legislative proposal has been presented on 16 December 2020, as a part of a package of measures to improve further the resilience and incident response capacities of public and private entities, competent authorities and the Union as a whole in the field of cybersecurity and critical infrastructure protection.

4 Reporting Obligations

After risk and security management, a largely regulated area in the field of data protection and cybersecurity concerns reporting obligations. The latter are closely linked to risk management as they relate to mandatory notification in certain cases of adverse consequences and to accountability for measures taken to prevent and tackle potential or present risks.

In this regard, the GDPR includes detailed provisions on notifications and reporting obligations in the event of security incidents (data breaches; Article 29 Working Party 2018). The Regulation requires controllers to report personal data security breaches to the competent supervisory authority without undue delay, unless they can demonstrate that the breach is unlikely to result in a risk to the rights and freedoms of data subjects (Article 33).

Controllers must also communicate the data breach to the data subject if such an event is likely to result in a high risk to the data subject's rights and freedoms (Article 34), but communication is not required if controllers have implemented appropriate prior or subsequent technical and organisational measures to render the personal data unintelligible (e.g., adequate encryption) or to exclude high risks to data subjects' rights and freedoms.

The EDPB has recently adopted guidelines (01/2021) on examples regarding data breach notification. These guidelines complement the WP29 (250/2017) guidance on data breach notification by introducing more practice-oriented guidance and recommendations. They aim to help data controllers in deciding how to handle data breaches and what factors to consider during risk assessment.

Finally, controllers must keep an internal register of incidents and personal data breaches, detailing the event and subsequent mitigation action taken (Article 33.5). This documentation will help controllers demonstrate their accountability and compliance with GDPR provisions.

Specific notification and reporting obligations are also set in the context of payment services by the PSD2 Directive, where payment institutions are required to provide specific information in their application for authorisation to operate on incident reporting mechanism which recognises the payment institution's notification obligations (Article 96).

In the event of a cyber incident or major physical security emergency, the companies must have a comprehensive incident response plan in place to manage the full range of tasks. This should include compliance with government reporting obligations and individual notification requirements.

Regarding notification obligations, payment service providers must report any major operational or security incident, without undue delay, to the competent authority in the provider's home Member State (Article 5). If the incident may have an impact on the financial interests of its payment service users, the provider must promptly inform users of the incident and any mitigation measures.

The national authority must report the incident to the European Banking Authority (EBA) and the European Central Bank (ECB) and, after assessing its importance, to the relevant authorities in the Member State and, if necessary, to other authorities. The ECB and EBA, together with the national authority, must assess the relevance of the incident to other Union and national authorities and notify them accordingly. The ECB will notify the members of the European System of Central Banks on any issues pertinent to the payments system.

Like the PSD2 Directive, also the eIDAS Regulation requires a range of responses in the event of a security breach, in particularly notification of the competent authority and a remediation plan to contain the spread of the breach (Article 10).

Firstly, where either the electronic identification scheme is breached or partly compromised in a manner that affects the reliability of cross-border authentication, the notifying Member State must, without delay, suspend or revoke that cross-border authentication or the compromised parts of it, and must inform the other Member States and the Commission.

Secondly, once the breach is remedied, the notifying Member State must re-establish cross-border authentication and promptly inform the other Member States and the Commission. If the breach is not remedied within three months of the suspension or revocation, the Member State must notify the other Member States and the Commission.

Regarding notification obligations, qualified and non-qualified trust service providers must, without undue delay, but in any event within 24 hours after having become aware of it, notify the supervisory body and, where applicable, other relevant bodies (e.g., the competent national body for information security or the data protection authority) of any breach of security or loss of integrity that has a significant impact on the trust service provided or on the personal data maintained.

Where the breach of security or loss of integrity is likely to adversely affect a natural or legal person to whom the trusted service has been provided, the trust service provider must also notify the natural or legal person of the breach of security or loss of integrity without undue delay. Where appropriate, in particular, if a breach of security or loss of integrity concerns two or more Member States, the notified supervisory body must inform the supervisory bodies in other Member States concerned and ENISA.

The notified supervisory body must inform the public or require the trust service provider to do so, where it has determined that disclosure of the breach of security or loss of integrity is in the public interest. Once a year, the supervisory body must provide ENISA with a summary of any security breach or loss of integrity notifications received from the trust service providers.

At a more general level, the Regulation requires Member States to fulfil two different types of reporting obligations. The first is the description of the electronic identification procedures in each Member State. Here the Commission publishes a list of the electronic identification schemes notified in the Official Journal of the European Union. The second is the notification of security incidents, in line with the general direction of recent European regulation.

Reporting obligations play a central role also in the NIS Directive, where Member States must ensure that operators of essential services promptly notify the competent authority or Computer Security Incident Response Team CSIRT of any incidents having a significant impact on the continuity of the services they provide. Notifications must include the information enabling the competent authority or CSIRT to determine the cross-border impact of the incident, if any. Notification must not entail greater liability for the notifying party. From this point of view, it is interesting to note that the new proposal of NIS directive has solved the possible conflict of the double sanction if it is ordered by both the Data Protection Authority and the competent National Authority (Articles 31 and 32 of the Proposal for a directive on measures for high common level of cybersecurity across the Union).

There are three criteria in determining the significance of the impact: number of users affected by the disruption of the essential service, duration of the incident, and geographical extent of the area affected.

On the question of notification, Member States must ensure that digital service providers notify the competent authority or CSIRT, without undue delay, of any incident having a substantial impact on the service they offer within the Union. Notifications must include the information enabling the competent authority or CSIRT to determine the cross-border impact of the incident, if any. Notification must not entail greater liability for the notifying party.

In determining whether the impact of an incident is substantial, the following elements must be taken into account: number of users affected by the incident, duration of the incident, geographical extent of the area affected, extent of the disruption to functioning of the service, and the extent of the impact on economic and societal activity.

5 Business Continuity, Disaster Recovery, and Resilience

Few and more general provisions are provided by the examined legal instruments with regard to Business Continuity, Disaster Recovery, and Resilience. These areas are mainly demanded to technical implementation of appropriate measures, considering the context-specific needs and potentially adverse situations.

In this regard, among the data security obligations, the GDPR also requires the adoption of measures to ensure data availability and recovery in case of loss or destruction resulting from a data breach (e.g., data backup and restore procedures; Article 32(1)(c)). More broadly, the GDPR aims to ensure the resilience (i.e., to the ability of the system to continue operating under adverse conditions) of the processing systems and services. Thus, controllers should take organisational measures to meet this requirement, such as Business Continuity plans, data restore procedures, effective cyber-resilience approaches, and Disaster Recovery plans. They are also required to adopt appropriate technological systems and tools to ensure business continuity (e.g., backup techniques and redundancy techniques; Article 32.1.b).

In the field of payment services, payment institutions are required by PSD2 Directive to provide a description of the business continuity arrangements adopted, clearly identifying the critical operations, the contingency plans and the procedures to regularly test and review the adequacy and efficiency of such plans.

For authorisation as a payment institution, an application shall be submitted to the competent authorities of the home Member State, together with a description of business continuity arrangements including clear identification of the critical operations, effective contingency plans and a procedure to regularly test and review the adequacy and efficiency of such plans.

Any provider should develop response and recovery plans, which should (i) focus on the impact on the operation of critical functions, processes, systems, transactions and interdependencies; (ii) be documented and made available to the business and support units and readily accessible in case of emergency; (iii) be updated in line with lessons learned from the tests, new risks identified and threats and changed recovery objectives and priorities.

eIDAS (Article 24.2) states that Trust Service Providers should have a well-defined business continuity plan (including disaster recovery plan and contingency plan) as part of its response planning.

This plan aims at describing all the arrangements foreseen by the Trust Service Providers, including processes and procedures, to recover as quickly as possible from any kind of major disruption regarding its network or its systems and continue to provide its services.

The business continuity plan should be maintained, tested and be subject of training. As part of its implementation, this plan will require the creation of two other plans: a disaster recovery plan and a contingency plan. Finally, according to the NIS Directive, both operators of essential services and digital service providers must ensure cyber-resilience, implementing business continuity management measures such as (i) cyber risk and vulnerability management; (ii) incident response team; (iii) alternative resources in the event of crisis; (iv) backup systems.

An excellent approach to achieve NIS compliance is to implement a cyber resilience programme that incorporates both robust cybersecurity defences appropriate to the risk and appropriate tools and systems for dealing with and reporting incidents efficiently.

International standards, such as ISO 27001 and ISO 27035, serve as ideal frameworks for achieving NIS Regulations compliance. In fact, Section 12 of the Regulations says that the measures DSPs adopt must take "compliance with international standards" into account. In addition, cyber incident response management, business continuity management and penetration testing can also help organisations achieve a heightened level of cyber resilience and facilitate compliance with the NIS Regulations.

6 Conclusions

In the previous sections, we have carried out a cross-cutting analysis of different legal instruments, not considered in their main objectives and foundational principles, but in their operational and technological implementation.

This analysis of the main legal sources making up the EU framework on data protection and cybersecurity in the business context leads us to conclude that the framework provided by the European legislator is not a patchwork, but a coordinated harmonious model. Similar technologies are required by differing regulations to address issues related to a common core based on four central pillars: risk-based approach, by-design approach, reporting obligations, and resilience.

The GDPR highlights the requirements that should be met by all service providers, while the other instruments contain provisions addressing specific sectors (essential services, banking, electronic communications and online transactions). The relationship between the various regulations is therefore from genus to species.

While the GDPR provides the general framework for business activities based on personal data, the NIS Directive increases the level of resilience of critical infrastructure against cybersecurity risks, the PSD2 Directive promotes the development of advanced payment instruments and increases the security of the system, and the eIDAS Regulation supports the integration of digital identity and trust services into application services. In a nutshell, the GDPR has a more general vision, while the other regulations have a more specific vision (i.e., critical infrastructure, advanced payment and digital signature), but all four regulations have common features with regard to technical and organizational measures to protect data.

This conclusion is bolstered by the obligations on data controllers defined in the GDPR (Article 24.1), which sees the nature, scope, context and purposes of processing, and the varying likelihood and severity for the rights and freedoms of natural persons, as parameters for the implementation of "appropriate technical and organisational measures" by controllers. The same notion of appropriateness appears again with reference to data processors needing to provide sufficient guarantees to implement appropriate technical and organisational measures (Article 28.1).

The GDPR does not define appropriateness but refers to it as a factor that entails a balancing test, as demonstrated by Recital No. 84 which points out a direct relationship between appropriate measures and risk assessment. A measure is therefore deemed appropriate if it addresses the risks involved in a given case of data processing.

Appropriateness is a contextual notion, dependent on the nature of the data processing, so that its meaning cannot be circumscribed by the GDPR, which is general in character. For application of provisions in context, we must turn to the three sector-specific instruments, the NIS Directive, PSD2 Directive and eIDAS regulation. Here the appropriateness of the required measures is framed in terms of the risks and available responses in the various contexts. GDPR (Article 28) places on service providers operating in these fields the general obligation to adopt more specific and tailored solutions.

This contextualisation of the GDPR obligations, however, does not compromise the security requirement. On the contrary, it clearly reveals the uniformity of approach of the EU legislator to the issues of data security and cybersecurity in the business environment. It highlights the existence of a common thread running through the entire framework which clearly revolves around a few key clusters of security measures and procedures (Table 1).

This coordinated analysis of the different legal sources has identified three main elements in the EU's regulatory approach: a balance between principles-based provisions and technical rules, a variety of technological solutions seen by law as crucial to achieving the EU objectives in data protection and data security, and a clustering of the entire legal framework around four core elements (risk assessment, by-design approach, reporting obligations, and resilience).

Finally, data-intensive technologies and datafication of social environment require effective and integrated implementation of existing provisions with a focus on procedural and technological requirements, not limited to the theoretical foundations of data protection and cybersecurity. In this regard, certification plays a key role in increasing trust and security in digital products and services, as confirmed by the recent EU Cybersecurity Act establishing an EU certification framework for ICT products, services and processes. Moreover, certification frameworks could be of great help in avoiding overlaps between obligations and ensuring a harmonised approach between the different legal frameworks described in this paper.

Though, at this stage of the implementation of the legal instruments examined, it is not possible to provide a fully integrated picture of the various obligations making up this common regulatory framework, as the GDPR is mainly a principles-based regulation and two of the other three instruments are directives. However, we have achieved two key objectives: (i) identify the common patterns of obligations deriving from the various instruments; (ii) highlight the relations between these obligations, including the technology-based organisational and security measures.

Table 1 Common core

Rules and principles	GDPR	PSD2	eIDAS	NIS
Risk assessment and security measures	• Risk analysis • DPIA • Technical and organisational measures • Records of technical and organisational security measures adopted • Vulnerability and penetration testing (e.g., vulnerability scanning; ethical hacking)	• Operational and security risk management framework • Control model • Physical security • Access control • Continuous monitoring and detection	• Use of authentication factors (Knowledge-based factors, possession-based factors, private keys) • Use of inherent factors	• Communication (email) risk assessment (domain keys identified mail, sender policy framework, domain-based message authentication, reporting and conformance) • Software management • Access control • Authentication factors
Data protection by design and by default	• Adoption of specific security requirements and procedures from the early stages of lifecycle development • Procedures to integrate data protection safeguards into processing activities • Specific technologies able to support privacy and data protection (PETs)	Secure technologies by design and by default (data minimisation, pseudonymisation, encryption, privacy-oriented users' profiles settings)	Use a catalogue of specific design patterns to develop solutions to known security problems	

(continued)

Table 1 (continued)

Rules and principles	GDPR	PSD2	eIDAS	NIS
Notifications, reporting obligations, and mitigation measures (data breaches)	• Appropriate procedures to establish immediately whether a personal data breach has taken place • Incident response plan • Data flow and log analysers • Tokenisation; encryption, etc	• Early warning indicators • Processes and organisational structures to ensure the consistent and integrated monitoring, handling and follow-up of operational or security incidents • Procedure for reporting	• Applications or open source software for quick and easy reporting • Technologies to classify annual incidents	• Mandatory report to the national agency in case of significant disruptions • Adopt alerting systems • Information collection on incidents • Provide information on security issues • Automatisation of notification systems
Business continuity, disaster recovery, and resilience	• Business continuity plan • Data restore procedures • Adoption of an effective "cyber-resilience" approach • Disaster recovery plan • Backup techniques • Technological measures to ensure business continuity	• Identify a range of different scenarios • Develop response and recovery plans	• Business impact analysis and threat analysis • Recovery time	• Cyber-resilience and business continuity • Cyber risk and vulnerability management • Incident response team • Alternative resources to use in case of crisis • Backup systems

This not only represents a basis for a future integrated compliance model but also a stepping stone for rule makers towards a more comprehensive technical and legal harmonisation of the different obligations in the national implementation of the framework in EU member states.

References

Article 29 Working Party (2017) Guidelines on data protection impact assessment (DPIA) and determining whether processing is "likely to result in a high risk" for the purposes of regulation 2016/679. https://ec.europa.eu/newsroom/article29/item-detail.cfm?item_id=611236

Article 29 Working Party (2018) Guidelines on personal data breach notification under regulation 2016/679. https://ec.europa.eu/newsroom/article29/item-detail.cfm?item_id=612052

Bygrave LA (2017) Data protection by design and by default: deciphering the EU's legislative requirements. Oslo Law Rev 4:105–120

ENISA (2016) Guidelines for SMEs on the security of personal data processing. https://www.enisa.europa.eu/publications/guidelines-for-smes-on-the-security-of-personal-data-processing

ENISA (2017a) Recommendations on European data protection certification. https://www.enisa.europa.eu/at_download/fullReport

ENISA (2017b) Handbook on security of personal data processing. https://www.enisa.europa.eu/publications/handbook-on-security-of-personal-data-processing

ENISA (2021) Data pseudonymisation: advanced techniques and use cases. https://www.enisa.europa.eu/publications/data-pseudonymisation-advanced-techniques-and-use-cases

European Data Protection Supervisor (2018) Opinion 5/2018. Preliminary opinion on privacy by design. https://edps.europa.eu/sites/edp/files/publication/18-05-31_preliminary_opinion_on_privacy_by_design_en_0.pdf

European Data Protection Board (2019) Guidelines 4/2019 on Article 25 data protection by design and by default. https://edpb.europa.eu/our-work-tools/public-consultations-art-704/2019/guidelines-42019-article-25-data-protection-design_en

Gellert R (2020) The risk-based approach to data protection. OUP, Oxford

Jasmontaite L et al (2018) Data protection by design and by default. Eur Data Protect Law Rev 4:168–199

Le Métayer D (2010) Privacy by design: a matter of choice. In: Gutwirth S, Poullet Y, De Hert P (eds) Data protection in a profiled world. Springer, Dordrecht, pp 323–334

Mantelero A, Vaciago G (2017) Legal aspects of information science, data science and big data. In: Dehmer M, Emmert-Streib F (ed) Frontiers in data science. CRC Press, Boca Raton, pp 1–46

Mantelero A (2018) AI and big data: a blueprint for a human rights, social and ethical impact assessment. Comput Law Secur Rev 34(4):754–772

Mantelero A (2021, forthcoming) Comment to Article 35 and 36. In: Cole M, Boehm F (eds) GDPR commentary. Edward Elgar Publishing, Cheltenham

Markopoulou D, Papakonstantinou V, de Hert P (2019) The new EU cybersecurity framework: the NIS directive, ENISA's role and the general data protection regulation. Comput Law Sec Rev 35(6):1–11

Raab C (2020) Information privacy, impact assessment, and the place of ethics. Comput Law Secur Rev 37:1–16

Copyright and Data Protection

Barbara Pasa

1 Objects of Exploration

Our smartphones, FitBit watch bands, all kinds of devices with sensors, from vehicles to furniture and dress; beacons in private and public spaces, social media, what ad we click on and when do we shop, measuring temperatures and fluids in industrial processes, etc.: we live in a data-driven ecosystem, which collects data as it goes along not only for testing performance of products but also for influencing our behaviour. Platform providers, data brokers and digital analytic firms are all participants in this data economy. 'Big Data' includes personal data that everyone is worried about, but a lot of data do not pertain to an identifiable individual, such as aggregated or de-identified data in different fields, business records, technical data, sensor data, market data, trend analysis, information on air pollution, climate data and demographic information. When combined, multiple datasets have a network effect: their volume and correlation allow predicting hurricanes as well as people's behaviours with a fair degree of accuracy. If that is the general outline, the software industry with platform providers, data brokers and digital analytic firms face common legal issues, from ownership to registration of Intellectual Property (IP) in software and other technology, from enforcement of copyrights and other IP rights against infringers to data protection restrictions.

This contribution, as evocated in its title, will delineate a fault line, which is becoming very central as data storage and data distribution technologies enable unprecedented levels of violation of the authors' rights and, at the same time, of intrusion into individuals' data protection. On the one hand, copyright law protects the rights of an author to their original works of authorship from the moment of the work's creation—personal touch in intellectual creations. On the other hand, data

B. Pasa (✉)
Università IUAV di Venezia, Venice, Italy
e-mail: bpasa@iuav.it

R. Senigaglia et al. (eds.), *Privacy and Data Protection in Software Services*,
Services and Business Process Reengineering,
https://doi.org/10.1007/978-981-16-3049-1_10

protection law safeguards the right of an individual to their personal data. The definition of 'original work' is independent of the medium that can be analogue or digital; similarly, the definition of 'personal data' is independent of whether data is stored in electronic or hard copy form.

Some caveats restrict the angle of observation. The survey on overlapping copyright and data protection laws will be realised prevalently within an EU-wide perspective, in comparison with the US law. It does not take into consideration the issue of free software and public domain software. Nor does it concern the discussion about legal ownership of informational aspects of personality or the potential harm to data subjects caused by accessing their personal data.

1.1 Protection Granted to Software

De facto, software companies are mainly concerned with strategies to protect and monetise IP rights.

In the first place, they are concerned about the protection granted to software, which can be different depending on the object to be protected. Protecting the code of a program can mean limited protection against copy attempts because the same program can be done with a different code function; thus, the copyright holder cannot assert copyright protection to prevent others from implementing the algorithm using a different code. European legislation introduced copyright protection for software with Directive 2009/24/EC on the legal protection of computer programs; the Court of Justice of the European Union (CJEU) interpreted the notion of 'software' broadly (cf. the case *SAS Institute* C-406/10) applying it to preparatory design material, machine code, source code and object code but not to the functionality of the computer program or to the format of data files. The format of data files might be protected as 'works' by copyright under Directive 2001/29/EC if they are their author's own intellectual creation. Thus, the program codes are protectable by copyright, in any form they are expressed, as long as they are the result of the author's intellectual creation. Copyright protection is automatic for any original work of authorship (in the US it must also be fixed in a tangible medium of expression). Concretely, the issue revolves around the thresholds for the originality of the work, and the question is whether a piece of data, number or any small work are protectable. On the other hand, patenting the program guarantees the programmer greater protection because a patent protects the functions of the software regardless of the code in which it is written. Patentability is admitted in Western legal systems, where an increased convergence occurs between the approaches used by the European Patent Office and the US regarding the threshold for patentability of computer-implemented inventions. Computer programs 'as such' are excluded from patent protection. However, software can be patented if presented as a 'method or a system of methods' or as a 'technical means implementing a method', which provides a solution to a technical problem in a certain technical domain. These can be computer-implemented inventions according to the European Patent Convention. Software that processes technical data (not abstract numerical entities), such as image processing, data compression, noise suppression, encoding/decoding, are potentially patentable software, as well as the control software of an industrial process, the software that controls an apparatus or a technical process (for instance, those used for assistive technology devices), memory management software for a PC or its peripherals or processors.

1.2 Protection Granted to Data

In the second place, these companies are concerned about the protection of data: just because they have access to data, it does not mean they can freely use and share data.

The copyright protection of databases is admitted, but it receives little protection in the US. This protection may be stronger in Europe by Directive 96/9/EC on the legal protection of databases. The protection depends on the selection, arrangements and organisation of information, which must be expressed in an original way, and the copyright protection does not extend to the content. In addition, the trade secret mechanism can be applied, but it is very hard to apply to external datasets. Directive (EU) 2016/943 on the protection of undisclosed know-how and business information (trade secrets) defines a trade secret as any information that is not generally known, has commercial value due to this secrecy and has been subject to reasonable steps to ensure it remains a secret. The 'trade secret' is broadly defined as to include nearly any data handled by a commercial entity (shopping habits and history of customers, customer lists and profiles, algorithms, information about customers' behaviour, creditworthiness, lifestyle, reliability, personalised marketing plans, etc.). There is no easy way to fulfil the data subject's right of access to data being processed when trade secrets protection conflicts with the right to data protection. Trade secrets substantially limit controllers' transparency obligations.

Data protection law also comes into play. Since the last couple of years, software and information service industries have been very concerned about strategies to access and use data. They have been complying with the regulations and guidelines regarding data control and disclosure and fair information practices. The range of data collected is wide: in the US most privacy statutes are sector-specific, and regulations differ for each type of data gathered as well as for the manner in which they are collected, granting to the web users certain rights against the collection of data, especially for marketing purposes (for instance, the collection of medical information requires HIPPA compliance), most contain exceptions allowing consent to information collection and processing. Platforms, in particular, have structured their data collection activities around presumptive consent and have configured their digital ecosystem and artefacts in ways that make user enrolment nearly automatic. Calls for accountability in Big Data analytics and algorithmic decision-making systems are thus motivated by a common concern. In the European Union, legal measures are of a more general nature: the Charter of Fundamental Rights of the EU (Articles 7, 8 and 17(2)), the General Data Protection Regulation 2016/679/EU (hereinafter GDPR), and the Directive 2019/1024/EU on open data and the re-use of public sector information, constitute the harmonised (not uniform) framework on data law. Their interpretations by national and supranational courts are intended to convey a fair balance between data protection and information privacy laws, IP laws, the rights to conduct a business and freedom of expression, but a renewed equilibrium is still along the way. The CJEU is finding its voice on this balancing (since the case *Promusicae*, C-275/06 and *Scarlet Extended*, C-70/10 to the *M.I.C.M.* case, C-597/19). Both in the US and EU, independent authorities supervise the administration's processing of personal data to ensure compliance with information privacy rules: the Federal Trade Commission (FTC) and the European Data Protection Supervisor (EDPS). They provide guidelines regarding fair practices based on their respective legislation but often fail to address all types of new situations.

2 Personal and Non-personal Data, Data Protection and Copyright

The algorithmic processes that manipulate the data function as 'information-age refineries'. In a process comparable to the milling and handling of grain to generate by-products optimised for different industrial productions, they convert data-based inputs into the forms best suited for their exploitation on an industrial scale. What are the legal boundaries when constructing algorithmic models? What constraints affect computer programmers in encoding algorithms in software?

On one side, scholarship on the relationship between law and the collection and processing of data and information typically considers such activities as raising problems of data protection and information privacy.

Data streams are artefacts designed for datafication. As some commentators pointed out, the processes of harvesting data resemble the harvesting of raw materials within an industrial system of agriculture. The collection of personal and non-personal information on an industrial scale inevitably adopts a curatorial stance regarding the items to be gathered. Strains of information are selected and cultivated precisely for their durability and commercial value within a set of information processing operations. As said, data can be both raw and cultivated, both real and artificial. Raw non-personal data is defined only by its representative characters (bits), and it allows to set data protection laws aside, since a supposed protection would represent a legal barrier to access through commons. Further, raw non-personal data is not protected by traditional IP law other than the trade secrets tool. Personal data, instead, is the object of both data privacy and copyright protection. Since such data often contains sensitive information, it raises several legal and political issues. There is a growing debate on whether the categories of personal, sensitive, anonymous and non-personal data reflect characteristics of data when it is collected, and whether they determine the level of protection granted to input data. These characteristics can, however, change over time and regions, as data is used for different purposes and is connoted by different symbolic meanings. For example, the German Supreme

Court has argued that there is no such thing as 'irrelevant data' when it comes to data protection law, as informational technologies might use it for purposes that affect the data subject. Also 'neutral data' can affect the right to information privacy, offer grounds for discrimination or cause other harms. The damage that can be done by data does not depend on any of the above-mentioned categories of data but rather on how data is used. Indeed depending on how data is used, the data per se might be protected or not by data law or copyright (or both).

Also the copy of personal data might be a subject matter of copyright protection. There are relatively few circumstances when this copy meets requirements of copyright protection, although, in principle, a copy of personal data (understood in a sense of the structure of data), created on the basis of the right of access recognised by the GDPR, could be protected by copyright.

The threshold necessary to qualify data per se or a copy of personal data as a 'copyrighted work' is its originality, e.g. being an author's (data controller/data subject, see below) own intellectual creation. As is well known, Directive 2009/24 on computer programs (Article 1), Directive 96/9 on databases (Article 3) and Directive 2006/116 on copyright for the photographic type of work (Article 6) are the only directives that provide details on the prerequisites for the protection of specific copyright-protected 'works', while Directive 2001/29 on copyright and related rights (Article 2(a)) left the definition of 'works' open. This legislative starting point has been modified by CJEU case law (since *Infopaq* C-5/08 on short sequences of text and *Painer* C-145/10 on simple photographs—work formats which often occur in social media): protectable works must fulfil the prerequisites of an author's own intellectual creation, e.g. the freedom of choice and personality of the author.

2.1 Personal Data 'Per se' and Copyright

Personal data 'per se' have similar nature as ideas, facts or mathematical concepts: they are excluded from copyright protection. In general, it is doubtful whether personal data could be considered as literary work in the sense of qualifying for copyright protection because they are usually insubstantial to be classified as a result of intellectual effort or usually have no degree of originality. Some authors consider syntactic information protectable, such as pictures or video showing data subjects, when it can be also qualified as semantic information because copyright protects 'expressions'.

In the case of pictures or videos of data subjects, data controllers have to determine whether the copyright holder is the data subject or someone else. If the picture was provided to the data controller by the data subject (author of the picture), the right of access to the picture has to be provided to the data subject. In cases where the picture was uploaded on social media by a third person and the data subject was tagged in the picture (data subject is not the author), the GDPR does not acknowledge if data controllers have the obligation to acquire IP rights from third parties in order to provide the right of access to data subjects.

2.2 Copy of Personal Data and Copyright

A different issue concerns the copy of personal data. The GDPR obliges data controllers to provide a copy, in writing or by electronic means, which is objectively meant for a 'physical' perception of the personal data undergoing processing to the data subject (Article 15(3)). The electronic or printed copy of the personal data undergoing processing could be classified as a property right, owned by the holder of the copy. Based on the circumstances, the holder might be the data controller or data subject.

Understanding the copy created on the basis of the right of access is a core requirement in order to consider the copy as a subject matter of copyright protection.

The GDPR, however, does not explicitly recognise the eligibility for copyright protection for the copy of personal data as such but only for the possible intellectual effort invested into the design of the personal data processing or software on which computer programs operate and from which the copy is generated (Articles 15(3)(4)). The copy, under the GDPR, is created by the data controller for the benefit of the individual with the aim to provide their control on personal data which are processed by the data controller. Such copy reflects almost no intellectual effort or original creativity of the data controller. Under the described circumstances, the copy would not qualify for copyright protection.

A copy of personal data might be a list of structured personal data, which provides information about the content of their life to data subjects (e.g. copy of personal data from a social platform wall). Such a copy could be qualified as a literary work by the data controller. Under certain circumstances, the data controller (usually a commercial entity, not a natural person) can be considered as the author or right holder of such copy because of their input in terms of creativity in finding, selecting, organising and presenting relevant personal data forming a summary of personal data for each data subject requesting access. Actual technology makes copyright protection of the copy more of a theoretical question. Technological development of the Internet enables the creation of copy without any human intervention, as computer-generated works. In the case where the automatic computer program generates a copy, the data controller could not claim authorship: a practical result that challenges the definition of (must-be-human) 'authorship' for copyright protection.

2.3 Copy of Personal Data and Copyright Protection of Software

A possible conflict between the copyright protection of software (Directive 2009/24 on computer program) in the context of protecting the rights of others (not data controllers) and the right of access of data subjects is explicitly mentioned in Recital 63 of the GDPR. Software protection should not be adversely affected by the right to access and obtain a copy of personal data.

There are at least two scenarios: (a) the right holder (author) of the computer program may decide not to provide copy because of their exclusive right of reproduction (Article 4(1)(a) Directive 2009/24). The easiest way to fulfil the right of access is to provide a copy of the algorithm concerned. Creating such a copy may qualify as an exclusive act of partial reproduction of a computer program. (b) Data controllers provide the electronic copy of personal data in a special format of software, which is not accessible to data subjects, and the copy cannot be opened by Microsoft Excel or Word, for example, installed in the majority of computers owned by data subjects. The easiest way to fulfil the right of access to the copy, in this case, is for data subjects to buy another computer program, which may be sold by data controllers themselves. From the data controllers' point of view, copyright protection of works of others and the possible infringement of their rights and freedoms (Article 15(4) GDPR) represent a simple argument on how to limit the quality and quantity of personal data provided on the basis of the right of access. As the data controllers are responsible for balancing conflicting rights, copyright law would probably prevail over the right of access; however, a neutral balancing should only be applied by the Data Protection Authorities.

2.4 Copy of Personal Data, Sui Generis Database Protection and Copyright

Personal data processed by a data controller may qualify for protection under Directive 96/9 on the databases, which provides two types of protection, copyright and *sui generis*. Databases are protected by copyright if the selection or the arrangement of content is the intellectual creation of the author (cf. CJEU since the case *Fixtures Marketing* C-444/02 and *Football Dataco* C-604/10 for the threshold of 'originality'), and by *sui generis,* which is a right that protects economic investment, qualitative or quantitative, of the maker of the database, a right that has been described as being similar to neighbouring rights of phonogram producers and film producers.

Each copy provided to the data subject from the data controller could be different in a sense of original organisation, structure, arrangement or format of the order and layout of personal data and other information. This freedom of the data controller to design the copy might represent their personal touch (Article 3(1) Directive 96/9 'copy as a database' and Article 2(5) Berne Convention 'copy as a compilation').

Data controllers usually collect and store personal data of all data subjects in data files in the form of databases. The collection is classified as a database when it is arranged in a systematic or methodical way and is individually accessible by electronic means. The data controller, who is processing personal data in the filing system, might be eligible for *sui generis* protection of their database. Creating a copy of personal data from a database protected by a *sui generis* right by the data controller (maker of database) is an extraction from the database, and—as said—the GDPR limits the right of access only in the case where the right to obtain the copy affects the rights and freedoms of others (Article 15(4)). It means that data controllers have the option of refusing full access to the processed personal data in a form of a copy because of the rights of others. Consequently, the lawfulness of the refusal is difficult to be verified by data subjects. This also means that data controllers are only in theory obliged to create (original) copy with personal data or information, but in practice data controllers may refuse to deliver the copy.

2.5 Right of Rectification and Copyright

Another possible conflict between IP and data protection has been identified in Canadian jurisdiction: translated to the EU context, it is based on the definition of 'processing' (Article 4(2) GDPR): 'any operation or set of operations which is performed on personal data or on sets of personal data, whether or not by automated means, such as collection, recording, organisation, structuring, storage, adaptation or alteration, retrieval, consultation, use, disclosure by transmission, dissemination or otherwise making available, alignment or combination, *restriction, erasure or destruction* [emphasis added]'.

The right of access and the subsequent right to obtain rectification of personal data may interfere with the copyright interests of the creators of the records because only creators have the right to make any change to their work, due to the author 'moral right' to the integrity of their work.

The right to data portability is not a right for an individual to misuse the information in a way that could be qualified as an unfair practice or that would constitute a violation of IP rights. That implies that in the EU, if a copy is considered to be protected by copyright, such protection could limit the right of the data subject to rectification or the right to erasure (known also as 'right to be forgotten') of personal data because data controllers have moral rights attributed to the copy, e.g. to object modification or derogation of their work.

3 Processing of Data, Data Analytics and Machine Learning

From a different perspective, the current discussion on machine learning and inferential analytics focuses broadly on two issues: (1) whether the training data used to construct a model (e.g. content uploaded or created by their users) is protected by IP laws and (2) whether the outcome of the algorithmic process can be protected under IP law.

3.1 *Data Mining for Training the Algorithmic Process*

Algorithmic processes might devise data collection strategies also on the basis of the obstacles posed by data protection and IP laws.

They refine non-personal data and user personal data to produce virtual representations that work to make human behaviours and preferences predictable and profitable in aggregate by producing tranches of data doubled with probabilistically determined purchasing and risk profiles. Their degrees of access to data are nuanced from 'open' (e.g. open data access approach, as in the case of the Directive 2019/1024 on open data) to 'controlled' (e.g. mandatory access approach, as in the GDPR, Articles 5, 6, 13, 15, 20 and in sector-specific laws—Regulation 1907/2006 (EC) on chemical data, Regulation (EU) 2018/858 on vehicle repair and maintenance of information, Directive (EU) 2019/944 in metering and consumption of data, etc.) and depend on how many 'actors' can access data.

As said above, in order for data protection rights to apply, data must be personal, suitable to identify the individual. The identifiability is fluid and can change over time, depending on technological progress. Companies can use anonymisation and pseudonymisation (the processing of personal data in such a way that the information can no longer be associated with a specific person affected without required additional information, as long as this additional information is stored separately and is subject to technical and organisational measures which ensure that it cannot be allocated to a specific or specifiable person), techniques to avoid identifiability. In such cases, data controllers are not required to comply with requests from data subjects under Articles 15–20 of the GDPR, if they are not in a position to identify the data subject, unless the data subject can provide additional information that allows the data to be re-identified. However, assessments are drawn from disparate, often non-intuitive features, and data sources increasingly drive decision-making about people.

3.1.1 Inferential Analytics

Inferential analytics can be used to infer people's preferences, weaknesses, sensitive information (sex, race, sexual orientation, health condition, etc.) and opinions (religious or political stances). These can form the basis for micro-targeting, nudging and manipulation by online advertisement. These inferences are based not only on data individuals have provided or that has been observed, but also on information derived or inferred from it, as well as from anonymous or third party data. Inferences drawn from anonymous and non-personal data still pose risks for data subjects. As a result, identifiability as a prerequisite to exercise individual rights creates a gap in the protection afforded to data subjects against inferential analytics. Third parties may have an interest in inferences and derived data due to their value or the costs involved, and may rely on techniques to create value (e.g. trade secrets).

The GDPR, the draft e-Privacy Regulation, Directive 2019/770 on certain aspects concerning contracts for the supply of digital content and digital services and Directive 2019/771 on certain aspects concerning contracts for the sale of goods attribute only limited rights over inferences to data subjects.

At the same time, Directive 2001/29 on copyright and related rights, Directive 2004/48 on the enforcement of intellectual property rights, Directive 2019/790 on copyright and related rights in the digital market and some provisions in the GDPR push to facilitate data mining and Big Data analytics by limiting data subjects' rights over their data. Directive 2016/943 on trade secrets also poses a barrier to accountability, as models, algorithms and inferences may fall under this framework of protection.

Only in the case where inferences would be interpreted as 'personal data', the GDPR could apply and allow data subjects to access, rectify, delete and object to them. Profiling and automated decision-making, which may include inferences, can already be contested (Article 22 GDPR). If data is anonymised or pseudonymised, however, data protection no longer applies. The legal prerequisites for anonymisation and pseudonymisation are not evident from the GDPR. In particular, the question arises as to whether re-identification can be carried out.

The so-called 'Article 29 Working Party' qualifies verifiable and unverifiable inferences as 'personal data' (for instance, the results of a medical analysis), but it leaves open whether the process behind that inference is a piece of 'personal data'. The CJEU's current jurisprudence is inconsistent.

3.1.2 Research Environment

When an algorithm is trained via data mining in a research environment, consent, license agreements and remuneration are not required to use data as inputs to train the model. Directive 2019/790 on copyright in the digital market complements the existing legal framework on copyright. Among other things, it governs the legal status of data mining concerned with research organisations such as universities, research institutes and cultural heritage institutions that use new technologies that 'enable the automated computational analysis of information in digital form, such as text, sounds, images or data, generally known as text and data mining. [...]' (Recital 8, Directive 2019/790).

For text and data mining activities in such research environments, Directive 2019/790 suggests mandatory exceptions to the copyright regime, precisely to the exclusive right of reproduction and to the right to prevent extraction from a database (foregoing a need for license agreements or remuneration). Further, Directive 96/9 on databases introduced an exception to the use of data to monitor trends. These exemptions are relevant when considered alongside the GDPR's exemptions (Articles 85 and 89), which already grant exemptions from most of the rights recognised in the GDPR (Articles 14, 15, 16, 18, 17(3)(d) and 21) for data controllers 'processing for archiving purposes in the public interest, scientific or historical research purposes or statistical purposes'. Universities, research institutes and cultural heritage institutions (including public–private partnerships) therefore are going to receive substantial exemptions in respect to data protection and IP requirements when training algorithms, and the GDPR in this case facilitates the creation of profiles and models built from inferences.

3.1.3 Statistical Purpose

When machine learning algorithms and statistical modelling are trained on datasets, the statistical purpose exemption applies, and data subjects are unlikely to have information privacy rights built from their personal data protection under the GDPR.

Unless statistical purposes exemptions apply, members of the training dataset will retain data protection rights over any personal data contained in the model and may be able to exercise rights in relation to it, but this will not equate to any control or rights over the model as a whole. Even if the model is built using personal data of a natural person (and in this case the statistical purposes exemptions no longer apply), the individual cannot object to its construction (and application) and has no rights over it. In other words, no control or rights over the model are likely to be granted at the current state of European legislation and case law. The facilitation of model constructions and the lack of individual rights, according to the GDPR, go parallel with the protection of training set and training parameters, the architecture and the entire machine learning system by database right, trade secrets and copyright law.

3.2 Data Generated Works Performed by the Algorithmic Process

A further issue is whether the outcome of the algorithmic processes can find protection in the overlapping area of IP law and data law. If Directive 2019/790 on copyright in the digital market and Directive 2001/29 could apply to work generated by algorithms, business interests and data subjects' rights remain polarised. The task of balancing rights requires weighing opposing interests and deciding which prevail, and need to be exercised on a case by case basis.

3.2.1 Copyright and Data Protection Overlapping in Social Media

Copyright and data protection laws overlap in certain systems of software development, precisely in social computing: it refers to the intersection of social behaviour and computational systems and to the mechanisms through which people interact with computational systems (how and why people contribute user-generated content and how to design systems that better enable them to do so).

This area of study, also known as social media law, is a new relevant subject for copyright and data protection scholars too: case law and academic research papers are still few. As is well known, social media is a virtual place where people express themselves and interact with other network members, a powerful source of information to be used in many fields. Social media such as Facebook, Instagram, YouTube, and Twitter that combine networking, microblogging, and commenting functions, comprise websites and apps that facilitate the creation, expression, and sharing of data and ideas among users who maintain a personal profile within the system, and then interact privately or publicly with other users within the social media platform, expand their connections by searching for other users or accepting connections suggested by the platforms, and may also leave the social networks and remove their connections. Posted texts, images, videos, audio, and other digital contents enjoy protection under copyright law if they are the author's own intellectual creation.

Most recently, platform businesses have begun to more directly acknowledge their pervasive manipulations of the information environment in the service of profit extraction and to recast those manipulations as inherently directed towards discovering 'scientific truths about human behaviour'. Platform-based media infrastructures, they argue, are information laboratories, in which providers of information services experiment to see which types of data are most useful and responsive to people's needs. The use of online shared contents from these multi-functional social media is now taken on by the much-debated Article 17, Directive 2019/790 on copyright in the digital market, which Member States shall bring into force by 7 June 2021.

3.2.2 Public Domain Issue

Contemporary practices of personal data processing could be qualified as a new type of public domain: a source of materials that are there for the taking and that are framed as inputs to particular types of productive activity.

The data extracted from individuals play an increasingly important role as material in the political economy of informational capitalism. Understood as processes of resource extraction, the activities of processing personal information mobilise different legal constructs, as seen above, from data protection to IP law.

A public domain is not a phenomenon that occurs naturally. It is a culturally situated way of understanding patterns of resource availability and their ownership. It is also a zone of legal privilege: the construct of a public domain both designates particular types of resources as available and suggests particular ways of putting them to work; it demarcates conduct as to which no one has a right to object and obscures the distributive politics in which patterns of appropriation are embedded—to misquote Wesley Newcomb Hohfeld.

The actual debate on the relative privilege consequent to the appropriation from public domain of personal and non-personal data and about 'disentitlements' thus require further epistemological work, whereas datafication and 'platformisation' industries have consolidated a *narrative* that links data processing with innovation, presents privacy and innovation as intractably opposed and perpetuates the romantic authorship ideal.

References

Afelayan MS (2018) Legal challenges of intellectual property and copyright protection of online and digital data in Nigeria. J Law Policy Glob 79:139–147

Arroyo Amayuelas E, Camara Lapuente S (2020) El derecho privado en el nuevo paradigma digital. Marcial Pons, Madrid

Barocas S, Selbst AD (2016) Big data's disparate impact. Calif Law Rev 104:671–732

Berger C (2018) Copyright law and data protection law. In: Khurshid A (ed) Social computing and the law. Cambridge University Press, Cambridge, pp 59–90

Cohen JE (2017) Law for the platform economy. U.C. Davis Law Rev 51:133–204. HeinOnline

Falce V (2018) Copyrights on data and competition policy in the digital single market strategy. Antitrust Public Policies 5:32–44

Falce V, Ghidini G, Olivieri G (2018) Informazione e big data tra innovazione e concorrenza. Giuffrè, Milano

Fia T (2020) An alternative to data ownership: managing access to non-personal data through the commons. Global Jurist. https://doi.org/10.1515/gj-2020-0034

Garcia R, Thaddeus A (2017) Social media law in a nutshell. School of Law Faculty Publications. https://ecommons.udayton.edu/law_fac_pub/21 Accessed 15 February 2021

Giovanella F (2017) Copyright and information privacy. Conflicting rights in balance. Edward Elgar, Northampton

Goldstein P, Hugenholtz B (2019) International copyright. Principles, law, and practice, 4th ed. Oxford University Press, Oxford

Grimaldi C (2021) A post for change: social media and the unethical dissemination of nonconsensual pornography. Hastings Commun Entertain Law J 43:109–134

Menell PS, Lemle MA (2020) Intellectual property in the new technological age, vol. I. Clause 8 Publishing, Berkeley

Moro Visconti R (2020) La valutazione dei "social network." Il Diritto Industriale 1:71–82

Pitruzzella G (2016) Big data, competition and privacy: a look from the antitrust perspective. Concorrenza e Mercato 1:15–28

Schlag P (2015) How to do things with Hohfeld. Law & Contemp Probs 1–2:185–234

Sobolciakova A (2018) Right of access under GDPR and copyright. Masaryk Univ J Law Technol 2:221–246

Sorkin D et al (2015) Legal problems in data management: IT and privacy at the forefront: big data: ownership, copyright, and protection. John Marshall J Inf Technol Privacy Law 4:565–585

Wachter S, Mittelstadt B (2019) A right to reasonable inferences: re-thinking data protection law in the age of big data and AI. Columbia Bus Law Rev 2:494–620

Wilkinson MA (2001) The copyright regime and data protection legislation. Law Publications 45. https://ir.lib.uwo.ca/lawpub/45. Accessed 19 February 2021

Zech H (2015) Information as Property. JIPITEC 6:192–197

Applicable Solutions

eHealth and Data

Carolina Perlingieri

1 eHealth and the Development of Multiple Technological Solutions to Improve Health and Medical Care

eHealth is the union of digital health and digital care. Digital health and care is the collective term used to refer to tools and services that use information and communication technologies (ICTs) that can improve prevention, diagnosis, treatment, monitoring and management of health and lifestyle. The use of digital tools and services, especially where it allows the transnational interoperability of health information, makes it possible to improve the health of citizens and enhance the quality and access to health care.

The eHealth plan provides technological solutions for different purposes such as greater informed patient participation (e-Patient through online health information); the locating of emergency calls (for example, AML, Advanced Mobile Location or eCall, which is mandatory on all newly approved passenger cars and light commercial vehicles); the improvement of human health and well-being (e.g. Digital well-being; Digital Therapies-Digital Therapeutics—DTx, a specialised software able to effectively and measurably guide a patient's progress towards improving his/her medical condition, for example, AmicoMed app; the m-Health, i.e. a mobile, portable and wearable device but also the use of Advanced Diagnostics assisted by Artificial Intelligence algorithms using Big Data); the enhancement of NHS services in terms of higher quality, efficiency, safety and access to care (for example, cup—Single Booking Centre; Electronic Health File, ESF; Electronic health record; telematic certificates of sickness; ePrescription, Electronic prescription; telemedicine (tele-diagnostics, remote assistance, telesurgery, etc.); and medical robotics).

C. Perlingieri (✉)
University of Naples «Federico II», Naples, Italy
e-mail: carolina.perlingieri@unina.it

R. Senigaglia et al. (eds.), *Privacy and Data Protection in Software Services*, Services and Business Process Reengineering,
https://doi.org/10.1007/978-981-16-3049-1_11

Therefore, the technological area of eHealth is very wide and under development. The examples above represent only some of the applications in the health sector whose functioning necessarily requires the collection of patients' health data for the purposes of the treatment of diseases, the improvement of the quality of life, the monitoring of patients, the research and improvement of diagnostics and the reduction of human error.

Consequently, the theme of the algorithmic processing of health data "collected directly or indirectly" and of the good produced by such processing and therefore of its use, which necessarily affects the data subjects for different reasons, becomes central.

2 Health Data and Their Necessary Qualification in Practice

In this respect, it is necessary to identify the relevant law.

Article 9 of the GDPR governs the processing of special categories of personal data "revealing racial or ethnic origin, political opinions, religious or philosophical beliefs, or trade union membership, and the processing of genetic data, biometric data for the purpose of uniquely identifying a natural person, data concerning health or data concerning a natural person's sex life or sexual orientation shall be prohibited".

The rules for processing these data are established by Article 9(2) of the GDPR under which its use is prohibited, unless there is the consent of the data subject or there is a super-individual requirement that legitimates the use of that data in the absence of authorisation.

In the presence of one of the conditions of lawfulness provided for by Article 9(2) of the GDPR, there are additional provisions for the processing including the request of explicit consent from the data subject (Article 9(2) GDPR); the mandatory adoption of a record of the processing activities (Article 30 GDPR); the need to carry out an impact assessment in the case of large-scale processing (Article 35 GDPR) with possible prior consultation of the Data Protection Authority under Article 36 of the GDPR; the appointment of a sectorally competent data protection officer (Article 37 GDPR); and a representative for processing carried out by a data controller or controller not established in the European Union (Article 27 GDPR).

Article 4(1)(15) GDPR defines "data concerning health" as "personal data related to the physical or mental health of a natural person, including the provision of health care services, which reveal information about his or her health status".

This concept is incorporated in Recital 35 of the GDPR, which specifies that "[p]ersonal data concerning health should include all data pertaining to the health status of a data subject which reveal information relating to the past, current or future physical or mental health status of the data subject. This includes information about the natural person collected in the course of the registration for, or the provision of, health care services as referred to in Directive 2011/24/EU of the European

Parliament and of the Council to that natural person; a number, symbol or particular assigned to a natural person to uniquely identify the natural person for health purposes; information derived from the testing or examination of a body part or bodily substance, including from genetic data and biological samples; and any information on, for example, a disease, disability, disease risk, medical history, clinical treatment or the physiological or biomedical state of the data subject independent of its source, for example from a physician or other health professional, a hospital, a medical device or an in vitro diagnostic test".

Previously, the WP29 document of 5 February 2015 "Annex—health data in apps and devices" established that health data are not only those generated in a professional medical context (medical data and raw sensor data) but also data related to physical health generated by devices or applications regardless of the qualification of the medical device when crossed with other data. Therefore, a qualification of the data is necessary in practice, so those data functionally suitable to reveal information related to a person's health can qualify as health data and compel the data controller to comply with the provisions laid down for the special categories of data relating to the health.

This approach is confirmed by a recent measure of the European Data Protection Board which, in providing guidelines on the processing of health data for scientific research against the COVID-19 emergency, specifies that information acquired on the basis of cross-references to other data revealing the health status or health risks can also be considered as health data (for example, the presumption that a particular person is exposed to a higher risk of heart attack based on repeated blood pressure measurements over a certain period); as well as information from self-assessment tests, in which interested parties answer questions related to their health (for example, describing symptomatology); information that becomes health data as a result of their use in a specific context (for example, information about a recent trip or staying in a COVID-19 region processed by a healthcare professional to make a diagnosis).

If the data controller plans to use or deduce health data, it will have to consider the effective use of the information and, based on the foreseeable results, establish the implementation of the regime provided for by Article 9 of the GDPR. Support to verify the suitability of the inferential processing to deduce health data is to be found in the impact assessment under Article 35 of the GDPR, which allows evaluating the possibility of obtaining information on the health of the data subject considering that "[t]he likelihood and severity of the risk to the rights and freedoms of the data subject should be determined by reference to the nature, scope, context and purposes of the processing" (Recital 76 GDPR).

3 Lawfulness of the Processing of Health Data and Mitigation of the Participatory Dimension of the Data Subject

It seems central, therefore, to analyse the different conditions of the legitimacy of the processing of health data.

Health data may be processed on the basis of the explicit consent of the data subject or due to some waivers in the case of the processing of personal data that have become public. This can be in the case of the processing necessary in order to protect the vital interests of a natural person; for the purposes of preventive or occupational medicine; for reasons of public interest in the area of public health; for archiving purposes in the public interest, scientific or historical research purposes or statistical purposes in accordance with Article 89(1) of the GDPR. Additionally, it can be to fulfil the obligations and exercise the specific rights of the data controller or data subject in labour law and social security or to establish, exercise or defend a right in court (*ex* Article 6(1) and Article 9 GDPR). Therefore, with regard to health data, the participatory dimension of the data subject is mitigated by the need to balance individual needs with demands relating to collective interests.

The areas relevant for the exemption, as remarked by the Italian Data Protection Authority, are linked to reasons of relevant public interest on the basis of Union or Member State law (Article 9(2)(g) GDPR), identified by Article 2-*sexies* Legislative Decree no. 196/2003 (as amended by Legislative Decree no. 101/2018). These are reasons of public interest in the area of public health, such as protecting against serious cross-border threats to health or ensuring high standards of quality and safety of healthcare and medical products or devices, on the basis of Union or Member State law which provides for suitable and specific measures to safeguard the rights and freedoms of the data subject, in particular, professional secrecy (Article 9(2)(i) and Recital 54 GDPR). Additionally, relevant for the exemption are the purposes of preventive or occupational medicine for the assessment of the working capacity of the employee, medical diagnosis, the provision of health or social care or treatment, or the management of health or social care systems and services on the basis of Union or Member State law or pursuant to a contract with a health professional (Article 9(2)(h) and Recital 53 GDPR; Article 75 Legislative Decree no. 196/2003 as amended by Legislative Decree no. 101/2018). It should also be considered that the exemption is justified in the event of processing necessary for archiving purposes in the public interest, in particular scientific ones, in line with Article 89(1) based on Union or Member State law which shall be proportionate to the aim pursued, respect the essence of the right to data protection, and provide for suitable and specific measures to safeguard the fundamental rights and the interests of the data subject, such as, in particular, minimisation measures, encryption and pseudonymisation techniques and specific arrangements for selective access to data (Article 9(2)(j) GDPR). The reference to Article 9(2) is also carried out by Article 2-*septies* and Article 75 of Legislative Decree no. 196/2003 establishing that appropriate safeguards must be taken with regard to the specific processing purposes and for making information to data subjects, along with any other measures necessary to guarantee the rights of data subjects.

The Italian Data Protection Authority also intervened on this delicate issue with Measure no. 146 of 2019 ("Requirements relating to the processing of special categories of data, in accordance with Article 21(1) of Legislative Decree no. 101/2018") which refers to the existence of "documented" reasons in the research project, particular or exceptional, for which informing the interested parties is impossible or involves a disproportionate effort or risk, making it impossible or seriously endangering the achievement of the research objectives.

The controller of medical and health information must not only adequately document the aforementioned reasons in the research project, but, where the research cannot achieve its objectives without the identification (even temporary) of the data subjects in the processing following the retrospective collection of the data, encryption or pseudonymisation techniques are adopted. Other solutions may also be implemented, which, given the volume of the data processed, the nature, object, context and purpose of the processing make them not directly attributable to the data subjects, allowing them to be identified only in case of need.

In addition, encryption techniques must also be adopted for the keeping of biological samples, and the conservation period following the conclusion of the study should be indicated in the research project after the conclusion of the study, at the end of which the aforementioned data and samples must be anonymised.

The adoption of such techniques, considering that genetic data can be extracted from the analysis of biological samples, makes it possible to ensure the protection, in particular, of those data "relating to the inherited or acquired genetic characteristics of a natural person which result from the analysis of a biological sample from the natural person".

4 The Rights of the Data Subject Relating to the Control of the Flow of Health Data: In Particular, the Right of Access to Data Relating to His State of Health

In this context, therefore, it should be considered that if the data subject can control the flow of data through the exercise of access, rectification and limitation rights to processing (Articles 15, 16, and 18 GDPR), it is difficult to oppose the processing (*ex* Article 21 GDPR) if such data are necessary for the execution of a task of public interest and in particular of scientific research.

The flow control by the data subjects is exercised mainly through the right to know that the data relating to their state of health results from medical examinations of all kinds (e.g. those resulting from medical records containing information relating to the diagnosis, results of examinations, opinions of doctors treating or any therapies or interventions carried out and to be informed about the purposes of the treatment, the type of data processed, any recipients of the data, the retention period and the existence of their automated use).

Under Recital 63 of the GDPR, "[e]very data subject should therefore have the right to know and obtain communication in particular with regard to the purposes for which the personal data are processed, where possible the period for which the personal data are processed, the recipients of the personal data, the logic involved in any automatic personal data processing and, at least when based on profiling, the consequences of such processing".

Health data, for example, may be subject to administrative access by third parties who claim a legal situation worthy of protection, since "the right to protection of personal data is not an absolute prerogative, but must be considered in the light of its social function and must be balanced with other fundamental rights" (Recital 4 GDPR) such as, for example, that of defence under Article 24 Italian Constitution, but always needing a balancing judgement under the circumstances of the case.

4.1 A Unique Form of Access to Data: The Electronic Health File Is One of the Mainstays of eHealth

A unique form of access to data used in the health sector is achieved through the Electronic Health File (ESF), which is one of the mainstays of eHealth aimed at achieving significant increases in the quality of services provided in the health sector and improvements with efficiency.

The ESF defined in Article 12 of Legislative Decree no. 179/2012, on "Additional urgent measures for the growth of the Nation" (converted, with modifications, into L. 17 December 2012, no. 221) is a collection of data and digital documents of a health and social health type generated by present and past clinical events concerning the patient, produced not only by health facilities but also by the data subject such as consents, requests for obscuration, revocation and any subsequent changes to the sharing of the ESF with other actors, and access policies with related metadata with reference to the lifecycle and conservation plan. The mentioned decree establishes the ESF, leaving its implementation to the Regions and Autonomous Provinces of Italy, in compliance with the current legislation on the protection of personal data, with the purposes listed in Article 12(2) such as (a) prevention, diagnosis, treatment and rehabilitation; (b) study and scientific research in the medical, biomedical and epidemiological area; (c) health planning, quality of care verification and health care assessment. Afterwards, the DPCM no. 178 on "Regulation on Electronic Health File" established its regulation and, with the Budget Law of 11 December 2016, no. 232, Article 12 of Legislative Decree no. 179/2012 was updated and defined as the National Infrastructure for the Interoperability (INI) of Regional Electronic Health Files. The latest amendments to Article 12 of Legislative Decree no. 179/2012 were introduced with Legislative Decree no. 34/2020 on "Urgent measures in the healthcare, support for work and the economy", as well as social policies related to the COVID-19 emergency in force since 19 May 2020.

4.2 *The Main Problematic Issues of the ESF. The Central Role of Metadata in the Effective Interoperability of ESF*

The purpose of such a digital tool is to be able to manage the entire life cycle of the documents since if it is no longer necessary to acquire the patient's consent for the insertion of documents and health data in the File, but it is necessary to acquire the consent of the patient both for the purposes of consultation (Article 12(5) Legislative Decree no. 179/2012) and for the obscuration of the data.

However, the main problematic issues of the ESF arise when the document is inserted by an operator from a Region or Autonomous Province other than the Healthcare Region. In fact, in such cases, while the collection of visibility and obscuration policies of the individual document is borne by the Region or Autonomous Province to which the Health Facility that the citizen has turned to belongs (Medical Care Provision Region), the management of the policies is borne by the ESF system of the Region or Autonomous Province of healthcare (Healthcare Region). Consequently, if effective interoperability of ESF data between national facilities and doctors is not ensured, the system could be highly critical.

For this purpose, metadata as a set of data associated with a computer document or computer file to identify and describe its context, content, structure and to ensure its storage, traceability and accessibility over time (e.g. patient identifier: tax code; type of document: hospital discharge letter) becomes particularly relevant.

In fact, the enhancement of metadata, related to the health documents of the patients of a certain Region (Healthcare Region or Medical Care Provision Region), are stored and indexed in a registry as the only database at the regional level. The use of metadata allows the identification of the repository, located within health facilities or centralised at the regional level, in order to access the document published in the ESF produced by the Healthcare Region or Medical Care Provision Region with the use of a unique and persistent storage identifier.

Furthermore, the enhancement of metadata allows the adequate storage of documents with advantages such as, avoiding unnecessary diagnostic investigations or duplication of tests that are still valid with obvious savings in time, therapeutic effectiveness and public expenditure; or enabling healthcare professionals to access all the information needed especially to deal with emergency cases.

4.3 The Implementation of the Operability of the ESF and the Mitigation of the Limitation of the Right to Data Portability. The Right to Erasure: Limitations. The Impact of the Right to Request for Obscuration of Health Data on the Right to Erasure

The implementation of the operability of the ESF could also allow the mitigation of the limitation of the right to data portability under Article 20 of the GDPR according to which the data subject has the right to receive the personal data concerning them in a format structured for common use and readable by an automatic device. Personal data are provided to a data controller who has the right to transmit such data to another data controller without hindrance only when the processing is based on consent and is carried out by automated means.

Indeed, Article 20 of the GDPR states that "that right shall not apply to processing necessary for the performance of a task carried out in the public interest or in the exercise of official authority vested in the controller" so that it appears to have been denied its applicability to the health area. With regard to data transferred for advertising, care, scientific research and similar purposes (Article 9(2)(h)(f)(i)(g)(j) GDPR), these cannot be portable due to the existence of advertising needs. This right is applied exclusively in processing based on consent or a contract, legal bases that may not be used within the health area. In addition, it could also lack the other requirement for the applicability of that right, which is the processing carried out by automated means.

Consequently, the full functionality of the ESF could guarantee the effective interoperability of health data allowing circulation between the different regional systems, private and public, and it could also allow data management between the different Member States in order to achieve an acceleration of health cooperation.

Furthermore, in healthcare, the right to erasure (*ex* Article 17 GDPR) is restricted in that it succumbs to the need for conservation if "it is necessary for exercising the right of freedom of expression and information; for compliance with a legal obligation (…) for the performance of a task carried out in the public interest or in the exercise of official authority vested in the controller; for reasons of public interest in the area of public health (…); for archiving purposes in the public interest, scientific or historical research purposes or statistical purposes (…); or for the establishment, exercise or defence of legal claims".

However, the difficulties of exercising the right of erasure could be mitigated by the exercise of another right provided for in Article 8 of DPCM no. 178/2015 under which "the patient has the right to request the obscuration of health data and documents both before feeding the ESF and after, ensuring the consultation only to the patient and the holders who produced them".

Therefore, if the control of health data flow meets a restriction on the limitation to erasure for the general requirements relating to the monitoring of public health and scientific research that still need to be evaluated in the light of the principle of proportionality and therefore of conservation needs that are not otherwise satisfactory, this storage must take place in compliance with the needs of the patient such as those of choice with respect to the subjects to whom access to the documents for which it is ensured over time, the characteristics of reliability, authenticity and integrity and legibility, pursuant to Article 44 Digital Administration Code (CAD).

5 The Role Played by Artificial Intelligence on the Processing of Personal Data. Problematic Aspects

Addressing the issue of digital health and data necessarily involves considering the role played by new technologies equipped with "self-learning" that process personal data for health purposes from early diagnosis to the development of advanced monitoring systems and support in operations, from the management of immense or shared databases to the study of personalised therapies.

The adoption of intelligent systems using health data raises some questions not so much in the assumptions of support for health activity, but when it replaces the doctor.

Article 22 of the GDPR establishes a general prohibition against decision-making based solely on automated processing "which produces legal effects concerning him or her or similarly significantly affects him or her". However, the prohibition does not operate when the automated decision (a) is necessary for entering into, or performing, a contract between the data subject and a data controller; (b) is authorised by Union or Member State law to which the controller is subject and which also lays down suitable measures to safeguard the data subject's rights and freedoms and legitimate interests; and (c) is based on the data subject's explicit consent.

Furthermore, with regard to healthcare in accordance with Article 22(4) of the GDPR, automated decisions using special categories of personal data are prohibited "unless Article 9 (2)(a) or (g) GDPR applies and specific measures are in force to safeguard the fundamental rights and the interests of the data subject".

Therefore, the automated processing of health data, in particular by Artificial Intelligence, is lawful only if there is the consent of the data subject (Article 9(2)(a) GDPR) or when AI technologies detect reasons of public interest on the basis of Union or Member State law which must be proportionate to the purpose pursued and provide for appropriate and specific measures to safeguard the fundamental rights and the interests of the data subject (Article 9(2)(g)GDPR).

5.1 *The Legal Basis of the Processing by Artificial Intelligence in the Case of Support to the Healthcare Professional and in the Case of Automated Processing of Health Data*

The application of Artificial Intelligence in healthcare could include both the support activities of a healthcare professional and fully automated processes.

In the first case, a legal basis of processing will be required: the consent for the processing of health data or the presence of one of the cases of Article 9(2)(b)(c)(d)(e)(f)(g)(h)(i)(j) of the GDPR.

In addition, the re-use of data, that is the use of data for further use from that for which they have been transferred, is also allowed, even in the absence of specific consent (Article 5(1)(b) and (c) GDPR) where the further processing of personal data takes place in compliance with the principle of limitation of the purpose. This means that it is not incompatible with the initial purposes for which the data had been collected, even when it takes place for storage purposes in the public interest, scientific or historical research, or for statistical purposes, as well as in compliance with the data minimisation procedure which must be adapted, relevant and limited to what is necessary in relation to the purposes for which they are treated.

Quite the opposite, where re-use is in practice incompatible with the use for which the data has been collected, the massive re-use of data requires express authorisation.

In the second case, falling within the scope of Article 22(4) of the GDPR, the Artificial Intelligence may process health data on the legal basis of the consent of the data subject or the existence of reasons of public interest on the basis of Union or Member State law which must be proportionate to the purpose pursued and provide for appropriate and specific measures to safeguard the fundamental rights and interests of the data subject (*ex* Article 9(2)(a) and (g)GDPR).

5.2 *The Silence of the Legislator, Either European or National, on the Delicate Issue of the Automated Processing of Health Data by Artificial Intelligence. The Re-use of Health Data by an Intelligent Health System. The Adoption of Specific Measures to Guarantee the Data Subject in the Case of Health Care and Development of Scientific Research. The Confirmation of the Centrality and Delicate Role of the Interpreter Who Will Have to Balance the Needs of Privacy and Data Protection with the Prevailing Requests for Health Protection*

If the thorny issue of the automated processing of health data by Artificial Intelligence must be referred to the Legislator, either European or National, its silence remains both in the recent proposal for a Regulation on European Data Governance of 25 November 2020 and in the Resolution containing recommendations to the Commission on a civil liability regime for Artificial Intelligence approved on 20 October 2020.

In fact, the proposal for a Regulation expresses the need to increase not only the use of data for purposes of general interest—including health care, the fight against climate change, the improvement of mobility, the facilitation of the production of official statistics, the improvement of the provision of public services, the support for scientific research—but also the collection, above all, of the altruism of data to encourage their European sharing. However, it does not regulate the methods of automated processing of the data collected for the aforementioned purposes.

The regulatory intervention in Recital 35 is limited to establishing that the collection is intended to "allow data analysis and machine learning" and Article 2(14) defines "secure processing environment" in order to carry out such operations in terms of "physical or virtual environment and organisational means of providing the opportunity to reuse data in a manner that allows the operator [...] to determine and supervise all data processing actions, including to display, storage, download, export of the data and calculation of derivative data through computational algorithms".

Furthermore, the Resolution in point G merely highlights the need to replace "the ambiguous term Artificial Intelligence" with the use of "automated decision-making" which "implies that a user initially delegates a decision, in part or in whole, to an entity using software or service; that entity in turn uses automated decision-making models to carry out an action on behalf of a user, or to inform the user's decisions in the course of an action".

In this direction, when the software processes on behalf of the healthcare professional, the data previously analysed to improve its ability to develop anamnesis then the re-use of data through Artificial Intelligence seems to fall back into the purpose of therapy.

When the re-use of data by an intelligent health system is not attributable to the sphere of care, it must be used to allow the development of scientific research in order to formulate innovative diagnostic or therapeutic hypotheses.

In this direction, Article 110-*bis*(4) of the Italian Privacy Code states that it is possible to re-use health data to the extent that it is carried out in connection with clinical activities, for research purposes, by the institutes of hospitalisation and care (public and private) due to the instrumental nature of the health care activity carried out by the aforementioned institutes with respect to research, in compliance with Article 89 of the GDPR. For this rule, the use of personal data in this context must be carried out in accordance with the principle of minimisation and with the adoption of specific measures to guarantee the data subject to include not only pseudonymisation, when the specific purpose pursued that requires archiving does not allow anonymisation, but also the use of technologies that allow the most secure provision of data, e.g. differential privacy, in order to facilitate the development of open health data, as can also be seen from the proposal for a Regulation on European Data Governance Act.

However, if the re-use of health data by intelligent systems is confirmed if there are specific guarantees for the data subject, it is still necessary to evaluate the specific use of the algorithm in relation to the purposes it pursues. Consequently, if it is possible and unnecessary to re-identify a patient, it is still necessary to verify the compatibility of the re-use with the specific legal basis. Therefore, the constant need for a hermeneutic activity that takes into account the results produced by the algorithmic processing of data, especially in health care, confirms the centrality and delicate role of the interpreter, who will have to balance the needs of privacy and data protection with the prevailing requests for health protection.

References

Article 29 Working Party (2013a) Opinion 2/2013 on apps on smart devices. https://ec.europa.eu/justice/article-29/documentation/opinion-recommendation/files/2013/wp202_en.pdf

Article 29 Working Party (2013b) Opinion 03/2013 on purpose limitation. https://ec.europa.eu/justice/article-29/documentation/opinion-recommendation/files/2013/wp203_en.pdf

Article 29 Working Party (2014) Opinion 05/2014 on anonymisation techniques. https://ec.europa.eu/justice/article-29/documentation/opinion-recommendation/files/2014/wp216_en.pdf

Article 29 Working Party (2015a) Opinion 1/2015 Mobile Health Reconciling technological innovation with data protection. https://eur-lex.europa.eu/legal-content/EN/ALL/?uri=CELEX:52015XX0716(01)

Article 29 Working Party (2015b) Annex—health data in apps and devices. https://ec.europa.eu/justice/article-29/documentation/other-document/files/2015/20150205_letter_art29wp_ec_health_data_after_plenary_annex_en.pdf

Article 29 Working Party (2017) Guidelines on the rights to data portability. http://ec.europa.eu/newsroom/document.cfm?doc_id=44099

Bolognini L, Pelino E (2019) Codice della disciplina privacy. Giuffrè Francis Lefebvre, Milano

Bolognini L, Pelino E, Bistolfi C (2016) Il Regolamento Privacy europeo. Commentario alla nuova disciplina sulla protezione dei dati personali. Giuffrè, Milano

Carro G, Masato S, Parla MD (2018) La privacy nella sanità. Giuffrè Francis Lefebvre, Milano

Chassang G (2017) The impact of the EU general data protection regulation on scientific research. https://ecancer.org/en/journal/article/709-the-impact-of-the-eu-general-data-protection-regulation-on-scientific-research. Accessed 15 Jan 2021
Ciancimino M (2020) Protezione e controllo dei dati in àmbito sanitario e intelligenza artificiale. Edizioni Scientifiche Italiane, Napoli
Cuffaro V, D'Orazio R, Ricciuto V (2019) I dati personali nel diritto europeo. Giappichelli, Torino
Davenport T, Kalakota R (2019) The potential for artificial intelligence in healthcare. Future Healthc J 6(2):94–98. https://doi.org/10.7861/futurehosp.6-2-94
European Data Protection Board (2020) Guidelines 03/2020 on the processing of health data for scientific research purposes in the context of the COVID-19 emergency. https://edpb.europa.eu/sites/edpb/files/files/file1/edpb_guidelines_202003_healthdatascientificresearchcovid19_en.pdf
Finocchiaro G (2019) Intelligenza artificiale e protezione dei dati personali. Giur. it., 1670-677
Finocchiaro G (2017) Il nuovo Regolamento europeo sulla privacy e sulla protezione dei dati personali. Zanichelli, Bologna
Garante per la protezione dei dati personali (2019) Chiarimenti sull'applicazione della disciplina per il trattamento dei dati relativi alla salute in ambito sanitario. https://www.garanteprivacy.it/home/docweb/-/docweb-display/docweb/9091942
Garante per la protezione dei dati personali (2019) Prescrizioni relative al trattamento di categorie particolari di dati, ai sensi dell'art. 21, comma 1 del d.lgs. 10 agosto 2018, n. 101. https://www.garanteprivacy.it/home/docweb/-/docweb-display/docweb/9124510
Granieri G (2017) Il trattamento di categorie particolari di dati personali nel Reg. UE 2016/679. Nuove leggi civ comm 1:165–190
Guarda P, Ducato R (2014) Profili giuridici dei Personal Health Records: l'autogestione dei dati sanitari da parte del paziente tra privacy e tutela della salute. Rivista Critica di Diritto Privato 3:389
Malgieri G, Comandé G (2017) Sensitive-by-distance: quasi-health data in the algorithmic era. Inf Commun Technol Law 32(1):118–140. https://doi.org/10.1080/13600834.2017.1335468
Marelli L, Lievevrouw E, Van Hoyweghen I (2020) Fit for purpose? The GDPR and the governance of European digital health. Policy Stud 1(5):468–487. https://doi.org/10.1080/01442872.2020.1724929
Mattsson T (2019) Digitalisation and artificial intelligence in European healthcare. Eur J. Health Law 26(4):285–288
Noorbakhsh-Sabet N, Zand R, Zhang Y, Abedi V (2019) Artificial intelligence transforms the future of health care. Am J Med 132(7):795–801. https://doi.org/10.1016/j.amjmed.2019.01.017
Perlingieri C (2020) Creazione e circolazione del bene prodotto dal trattamento algoritmico dei dati. In: Atti del 14° Convegno Nazionale SISDiC. Il trattamento algoritmico dei dati tra etica, diritto ed economia. Edizioni Scientifiche Italiane, Napoli, pp 177–196
Perlingieri C (2019) Data as the object of a contract and contract epistemology. Ital Law J 5(2):613–629. https://doi.org/10.23815/2421-2156.ITALJ
Perlingieri P (2020) Relazione conclusiva. In: Atti del 14° Convegno Nazionale SISDiC Il trattamento algoritmico dei dati tra etica, diritto ed economia. Edizioni Scientifiche Italiane, Napoli, pp 379–394
Pizzetti F (2018) Intelligenza artificiale, protezione dei dati personali e regolazione. Giappichelli, Torino
Pizzetti F (2016) Privacy e il diritto europeo alla protezione dei dati personali. Dalla Direttiva 95/46 al nuovo Regolamento europeo. Giappichelli, Torino
Ruffolo U (2020) Intelligenza artificiale. Il diritto, i diritti, l'etica. Giuffrè Francis Lefebvre, Milano
Schönberger D (2019) Artificial Intelligence in healthcare: a critical analysis of the legal and ethical implications. Int J Law Inf Technol 27(2):171–203. https://doi.org/10.1093/ijlit/eaz004
Sica S, Sabatino BM (2020) Algoritmi e salute. In: Ferrari GF (ed) Smart city. Evoluzione di un'idea. Mimesis, Milano, pp 553–580

Location Data and Privacy

Alberto Maria Gambino and Davide Tuzzolino

1 About Location Data

Location data are specific kind of data that encompass information referring to a geographic or geospatial position. The process of geolocation implies the activity of correlating certain information regarding, *inter alia*, altitude, latitude and longitude to identify a specific point on Earth. A further step is represented by the ability to locate the position of devices in space, dynamically.

Location data are collected through a particular device, such as a mobile phone, tablet or wearable (i.e. fitness tracker). These devices store and send geolocation using different methods.

The best-known system is, of course, the Global Positioning System (henceforth GPS) which consists of a constellation of satellites launched, at first, by the United States of America to implement the positioning and navigating of military systems. It facilitates—using a geometric method called three-dimensional trilateration—the calculation of its exact location on Earth by signals from the satellites. Such a system provides accurate positioning. However, it is slow and functions only outdoors.

Other methods worthy of mention are:

- Wi-Fi access points, which allow devices to scan surrounding access points, from which the device detects an ID. It is important to emphasise that devices do not need to connect to the access point, since the mere detection in a passing scanning mode, allows the detection and collection of the data from the devices.
- Cell Phone Tracking, which uses radio receiving for mobile services to provide the mobile phone communication service. The area covered by telecommunication

A. M. Gambino (✉) · D. Tuzzolino
European University of Rome, Rome, Italy
e-mail: alberto.gambino@unier.it

D. Tuzzolino
e-mail: davide.tuzzolino@unier.it

R. Senigaglia et al. (eds.), *Privacy and Data Protection in Software Services*,
Services and Business Process Reengineering,
https://doi.org/10.1007/978-981-16-3049-1_12

operators is divided into cells, which enables the cell network area connection to be known, provided the owner of the device performs the Internet connection themself. This way the telecom operator can estimate, albeit approximately, the position of the device.

- Bluetooth Beacon transmitters, which are contained in objects such as key rings, send a unique identifier—used in helping to determine the object's location—to a compatible app installed on a smart device.

Furthermore, modern devices can merge data detected from the methods mentioned above, improving the accuracy of location through other tools.

Recently, location data have been used to provide different services and to perform the core purpose of certain mobile applications (henceforth apps), which are software developed to run on a mobile device. Location data could reveal private information that, due to the ubiquity of new generation devices, could relate to health data, financial data, consumer behavioural data, personal habits and the religious beliefs of its owner, thereby inferring a wide range of information about the life of the individual. The capacity of a very vast range of devices to process personal and non-personal data increasingly gives rise to many concerns about the processing at issue, both from a legal and ethical perspective.

The use of apps has become one of the main reasons for the disclosure of personal data and their processing. The increase in the collection of location data during the last twenty-five years witnesses the meteoric spread of new devices among the population, despite their age, occupation and interests. This is in part because apps are easy to obtain, thanks to the mobile application store (henceforth app store), a platform where users may browse through apps, divided into categories, and download them by a mere "touch".

In fact, some devices could be defined as a universal and unique tool, able to shape themselves on the basis of the purposes of the user. There are several categories of apps - easily downloadable from app stores installed in the device - regarding every category of the interests of a person, such as jobs, entertainment, social networks, leisure, news, sports, and fitness.

The access to different kinds of personal data and metadata is dictated by the characteristics of the hardware inside the device, which increases its potentiality in terms of picking up personal data. Every mobile phone, for instance, is equipped with a series of sensors (microphone, camera, infrared, GPS, Bluetooth, accelerometer, Wi-Fi, fingerprint sensor, etc.) that have the ability to retrieve personal information about the owner of the device. Hardware features also increase the potentiality of the operating system (henceforth OS) which is usually designed to fit perfectly with the characteristics of the device. The OS communicates with those sensors employing an interface that acts as an intermediary for the flow of data between services. Such an intermediary is named the Application Programming Interface (henceforth API). Moreover, apps use APIs to read information from sensors and interact between them and the OS.

Almost without exception, the hardware of new devices encompasses GPS chips, Wi-Fi or Bluetooth connections, which in many cases both the OS and downloaded apps exploit, thus collecting location data to ensure the full utilisation of functionality of the devices.

Legal issues related to the collection of location data via apps are always in the spotlight, due to the unpredictable and continuous evolution of matter. The Covid-19 pandemic raises the issue of the geolocation pertaining to the contact tracing method. But before diving into this topic, it would be appropriate to briefly outline the legal framework.

2 European Framework

Due to their revelatory nature, location data should fall under the category of personal data. The General Data Protection Regulation of the European Union n. 2016/679 (henceforth GDPR) is the main legal reference point for the processing of personal data. Following that, the Data Protection Directive 95/46/EC was repealed. The development and functioning of apps should be subject to the principles of GDPR.

By focusing the attention on location data, it is made clear from the definition of personal data provided by the GDPR that they can reveal personal information, thus falling under the scope of such Regulation. Indeed, according to Article 4(1) of GDPR, personal data means: "any information relating to an identified or identifiable person (data subject); an identifiable natural person is one who can be identified, directly or indirectly, in particular by reference to an identifier such as a name, an identification number, location data, an online identifier or to one or more factors specific to the physical, physiological, genetic, mental, economic, cultural or social identity of that natural person".

This means that the processing of location data shall be GDPR compliant insofar as they reveal personal information. The field of applying the GDPR encompasses also metadata. These are data that provide information about other data. For instance, new cameras enrich the photo with information about the author, the model of the camera, the date and hour of the photo, the settings of the camera and the GPS coordinates (so-called geotagging). The extra information is easily retrievable from the photo and, in the case of GPS data, reveals personal data inferable from the nexus between the place and the photo.

However, the European legal framework about mobile apps is wider and encompasses the e-Privacy Directive n. 2002/58/EC, as amended by Directive 2009/136/EC, concerning the protection of personal data in the electronic communications sector. The Directive is now subject to be updated through the new e-Privacy Regulation, which is currently being examined by the European Parliament and Council. The draft of the proposal was submitted by the European Commission in 2017.

The GDPR is, in this way, complemented and particularised by a specific discipline. The e-Privacy Directive establishes rules to ensure privacy and personal data protection in the field mentioned above. The matters which are outside of the scope of the Directive, but which concern the processing of personal data, are covered by the GDPR.

The interplay between the e-Privacy Directive and the GDPR has been recently analysed in an Opinion of the European Data Protection Board (henceforth EDPB). According to the EDPB, a series of provisions of the e-Privacy Directive do particularise the GDPR's one. Indeed, between these two laws, the principle *lex specialis derogat legi generali* is implemented (cf. Recital 173 of GDPR and CJEU, Joined Cases T-60/06 RENV II and T-62/06 RENV II of 22 April 2016, at paragraph 81). In this manner, the derogation of the general rule provided by the GDPR shall be carried out insofar as the e-Privacy Directive contains specific rules, with the caveat that the mentioned interplay involves the national transposition law of the Directive. Certainly, the latter sets out only the goal that the Member States shall achieve and require an internal law to reach these goals.

In the context of the e-Privacy Directive, location data, according to Article 2(c), "means any data processed in an electronic communications network or by an electronic communications service, indicating the geographic position of the terminal equipment of a user of a publicly available electronic communications service".

Moreover, Recital 14 specifies that such data "may refer to the latitude, longitude and altitude to the user's terminal equipment, to the direction of travel, to the level of accuracy of the location information, to the identification of the network cell in which the terminal equipment is located at a certain point in time and to the time the location information was recorded".

The Directive takes into account the idiosyncratic feature of electronic communications of revealing personal information about users involved. The proposal of the e-Privacy Regulation explains the concept of metadata, "data processed in an electronic communications network for the purposes of transmitting, distributing or exchanging electronic communications content; including data used to trace and identify the source and destination of a communication, data on the location of the device generated in the context of providing electronic communications services, and the date, time, duration and the type of communication".

Considering the reference scenario, the general principle of location data protection applied to processing via apps may be defined.

In the context of the processing of location data by a mobile device, the role of the data controller - who determines the purposes and means of the processing - could be interpreted by different players: (a) the developer of the OS could be a data controller if such software collects location data to improve services, (b) the provider of apps which process location data (i.e. maps, weather service, food delivery service) once installed on the device or accessed through a browser, and (c) data controllers of the geolocation infrastructure such as telecom operators or Wi-Fi access points. In addition, any other party that carries out further processing of location data collected shall be considered a data controller since they define the purposes and means of such operations.

Location data are collected and processed when the device allows the interaction between the sensor and the apps, or OS, even while such data are processed on the device or via the web.

The collection of such data shall be compliant with the general principles established by the GDPR that could be summarised as follows: (a) personal data shall be processed with lawfulness, fairness and transparency; (b) such data shall be collected for specified, explicit and legitimate purposes and not further processed in a way incompatible with those purposes; (c) processed accurately, adequately and limitedly to the purposes; (d) kept in a form which permits identification of data subjects for no longer than is necessary; and (e) safeguarded against unauthorised or unlawful processing and accidental loss, destruction and damage, adopting appropriate technical or organisational measures.

According to the above, apps could collect only those data that are strictly necessary to perform the functionality identified and planned. Any further and incompatible processing shall be considered excessive and thus unlawful.

The Regulation also retains the conditions for the processing of sensitive personal data, which are named special categories of personal data by Article 9 of GDPR. Those data are related to racial or ethnic origin, political opinions, religious or philosophical beliefs, genetic biometric and health data or data concerning a natural person's sex life or sexual orientation. Location data may reveal information that are inherent to those included in the special category of personal data in an inferntial way, thus their processing requires particular attention in point of proportionality and minimisation, which should lead to the necessity, for the controller, to perform a Data Protection Impact Assessment (cf. the California Privacy Rights Act that considers a consumer's precise geographic location as sensitive personal information and the jurisprudence of the United States Supreme Court on the point).

Thus, both the operative system and apps of devices pay attention to the issues related to pre-emptive and explicit consent.

Under Article 6 of GDPR, the legal ground of the processing could be the consent of the data subject given through the acceptance of the conditions of the processing of personal data implied by the service; even though the processing may be based on another condition (legal obligation, legitimate interest, etc). The consent shall be collected before the processing itself and shall be revoked at any time. It shall be specific, expressed and freely given. The prior consent is also the basis for the processing of location data collected from electronic communication providers, which shall be processed within the meaning of Articles 6 and 9 of the e-Privacy Directive. And, under no circumstances could the device lawfully transmit location data only based on the acceptance of general terms and conditions, as well as through the setup of geolocation reception as a default setting, without the intervention of the user. The processing needs that this latter makes is indeed an informed choice and, therefore, apps shall ensure granular consent specifying any category of personal data that will be processed and every purpose of the personal data processing at issue.

If those purposes change, the data controller shall inform any data subject promptly. Moreover, for some categories of personal data, such as location data, users shall be informed of the processing contextually to the single collection, by means of a specific icon on the screen of the device. The users thus are warned that the device is reading the details of the location, as long as this service is active.

The withdrawal of the consent shall be as easy as possible and without negative consequences for the users.

Special attention shall also be given to the consent of children. In this case, such consent shall be provided by parents or legal representatives before collecting and processing personal data.

In this field, the processing of personal data performed by some apps developed for parental supervision has attracted the attention of the Article 29 Working Party. The latter wrote in its Opinion 2/2009: "It should never be the case that, for reasons of security, children are confronted with over surveillance that would reduce their autonomy. In this context, a balance has to be found between the protection of the intimacy and privacy of children and their security".

Data controllers shall ensure that the data subject is fully informed on how to exercise his or her rights under the GDPR, such as the right to access, to rectify, to erasure and the new right to data portability.

The principles and rules of the GDPR are applicable in the case of the processing of location data as long as they are not anonymised. Indeed, the Regulation shall apply to information regarding an identified or identifiable natural person (cf. Recital 26 of GDPR). Thus, anonymised data are non-personal data and therefore must fall outside the scope of the GDPR and their flow is under the Regulation (EU) 2018/1807 on non-personal data (cf. Recital 9 of GDPR), whereas pseudonymised data are a full-fledged personal data.

It would be, therefore, useful to recap briefly the differences between anonymisation and pseudonymisation process.

The former is a technique through which the link between a personal data and an identified or identifiable data subject is removed, while the pseudonimisation process maintains a certain margin of reversibility. Location data shall be subject to the former procedure but only if aggregated in a dataset (and sometimes the aggregation still reveals information, as happened a few years ago when a fitness app - which retrieved aggregated location information through a heat map - revealed coordinates of one military classified location). Anonymisation of location data, according to the EDPB, is more difficult to perform. Moreover, some mobility traces reduce the possibility that the procedure succeeds, and the anonymisation of a single data pattern about the location of a person during a significant period contributes to increasing risks.

Pseudonymisation is an elaboration of personal data in a manner that prevents the attribution of data to a specific person without the use of additional information, which is kept separately. This procedure is a measure incentivised by the GDPR to ensure security in data protection, and concerning location data is always desirable.

3 Some Selected Cases

Location data are often aimed at purposes that go further in the revealing of the location itself. For instance, location data are used to have more control over properties and discourage theft, to improve the contextualisation of some services and to allow the further development of new technologies through those data (one thinks of cruise control and its evolutionary path towards autopilot). Furthermore, location data has recently been at the centre of the debate on prevention and containment measures concerning the spread of the global pandemic.

The following gives some examples of how governments, legislators, authorities and expert groups have adopted different approaches to such various situations, making an appropriate distinction depending on the context of location data usage.

3.1 Location of Employees

The use of geolocation of employees brings issues into question that found responses at the European Union level and in the domestic legislation of several Member States. In this field, WP29 found the legal basis in the legitimate interest of the employer, who is the data controller of the processing of personal data. In this context, the processing carried out by the employers arose from the necessity to supervise workers while balancing the opposing interests of the two parties. While, talking about consent, the WP29, in its Opinion 8/2001 on the processing of personal data in the employment context, pointed out that: "where consent is required from a worker, and there is a real or potential relevant prejudice that arises from not consenting, the consent is not valid in terms of satisfying either Article 7 or Article 8 as it is not freely given. If it is not possible for the worker to refuse it is not consent. (…) An area of difficulty is where the giving of consent is a condition of employment. The worker is, in theory, able to refuse consent, but the consequence may be the loss of a job opportunity. In such circumstances, consent is not freely given and is therefore not valid".

For instance, in the light of those premises, Italy has laid out specific provisions when it comes to using technology to locate workers. The current Article 4 of the Italian Workers' statute ("*Statuto dei lavoratori*"—Law n. 300/1970) allows remote control of employees only if compatible with the protection of their freedom and dignity, but, in any case, avoiding his or her massive, prolonged, and indiscriminate control (allowing, for instance, the shutdown of the tracker). The Statute indicates that technological tools shall be used for organisational and production needs, for workplace safety and the protection of company assets and can be installed after the collective agreement entered into by the unitary union representative or by company union representatives. Hence, vehicles are not generally considered as devices of control of working activity, and it shall not be excluded the respect to the GDPR principles on personal data processing in terms of providing to the data subject (the worker) information about the processing of personal data under Article 13 of the GDPR, concerning privacy by design and privacy by default approach (cf. Recital 78 of the GDPR), without undervaluing the appropriate distinctions, for instance, between the control of a company car and a commercial vehicle in a perspective of proportionality.

In France, following the same line, the French Data Protection Authority (Commission Nationale de l'Informatique et des Libertés or CNIL, in Délibération

2015-16) considers the use of geolocation of employees only for specific applications such as (a) to control services strictly related to the vehicle usage, (b) to ensure the security of workers and goods, (c) to check working hours. Geolocation is considered to be an intrusive measure that requires a prior Data Protection Impact Assessment (DPIA). Indeed, the CNIL excludes the use of geolocation to control working times, speed limits of a company vehicle, the permanent control of workers and the collection of personal data outside working hours.

Other national data protection authorities - such as the Hungarian Data Protection Authority (Nemzeti Adatvédelmi és Információszabadság Hatóság, or NAIH, in the opinion of October 2016) and the Portuguese Data Protection Authority (Comissão Nacional de Proteção de Dados, or CNPD) - issued guidelines on the use of location data to monitor employees consistent with the position expressed above and shared by the WP29 and the other Member States, allowing the processing of location data in such a context mainly for logistical purposes.

3.2 Smart Vehicles

New models of vehicles are connected via electronic communication networks, road infrastructure and telecommunications operators. This way cars may provide advanced and interactive services, to increase the range of accessories such as driving assistance, autopilot, vehicle condition, dynamic mapping with real-time updates on road conditions and traffic information and other entertainment services. Hence, smart vehicles generate and collect a vast amount of personal data about drivers and passengers which leads to consider the issues related to the collection, processing and storing personal data performed by smart cars.

The EDPB, indeed, recently issued a Guideline on processing personal data in the context of connected vehicles and mobility-related applications. The Board has identified three categories of personal data that deserve particular attention: location data, biometric data and data that could reveal offences or traffic violations.

In this context, location data processed are considered particularly revealing on the habits and interests of the data subject. Indeed, the itinerary covered provides good inferable information about places usually visited by the drivers, such as their home, workplace and where he or she used to spend free time, offering the opportunity to evaluate and profile some intimate aspects of the data subject.

In this case, data controllers could be identified, *inter alia*, in the vehicle producers, equipment manufacturers and service providers. The data controller - in obtaining the consent of the data subject - shall emphasise the highlights of the pattern of processing of personal data, making the data subject aware, for instance of the frequency of collection of tracking data, the possibility to shut down the tracking system and, depending on the service provided, in which cases location data could be transmitted to third parties (e.g. in case of declaration of theft), identifying those parties explicitly.

In collecting location data, data controllers shall pursue all the principles of the GDPR. They shall modulate the frequency of access and the level of detail of those data according to the purposes pursued: for instance, the collection of location data aimed to retrieve weather information should be carried out less frequently (and with less precision) than the one performed by the navigation system to provide the best itinerary. The purpose shall also influence the length of the storage of personal data, which shall be limited to a certain time, beyond which the data shall be deleted (data minimisation principle).

The vehicle, as far as possible, shall process personal data internally without sending them outside. The technologies shall be designed, in general, to minimise the privacy risk and respect the obligations of privacy by design and by default under Article 25 of the GDPR, which implies not only to project the architecture of the vehicle's system according to the Regulation, but also to make the other communicating infrastructures - which should receive and process location data unless absolutely necessary, for the shortest possible time, ensuring anyway adequate security measures - compliant too.

3.3 Contact Tracing

The spread of Covid-19 necessitated the need to develop a technological tool able to face this phenomenon. For example, facial recognition, instead of iris scanning and fingerprinting, and the use of a thermo-scanner reduced opportunities for physical contact. Some countries have adopted those tools, such as the People's Republic of China and the Russian Federation.

Moreover, health apps have played, and are still playing, a key role. From the mobile apps for the aim of prevention, diagnosis and remote medical assistance to the apps used for tracking devices, the use of those apps has been made useful to deal with the crisis. Contact tracing has always been an important instrument for the prevention and control of the spread of communicable diseases. Digital Contact Tracing (henceforth DCT), using database access and tracking technologies - such as GPS, Wi-Fi or Bluetooth mentioned above - can collect data to infer information about the simultaneous presence of individuals in a certain place.

Many countries have invested in the development of DCT apps. For instance - apart from the EU - China, South Korea and New Zealand relied on DCT, albeit with different approaches.

The fact that any of these has been introduced in different legal backgrounds cannot be overlooked.

However, these apps have raised doubts and concerns, almost everywhere, about privacy and personal data protection implications in general. Undeniably, the implementation of principles - like proportionality of collection, prior consent, minimisation, transparency, anonymity and accountability - were almost everywhere recommended.

This is a broad topic, but for this paper, it shall be limited to the scope of location data.

Location data may offer an overall view of the movement of persons. Such information can be useful both in the assessment of containment measures, such as lockdown and reduction of travel, and in the context of DCT.

In the field of DCT, proximity data should be distinguishable from geolocation data. Proximity data are generated by the exchange of Bluetooth Low Energy (henceforth BLE) among the closest devices, and they reveal relative position. Instead, geolocation data exploits information about geographical coordinates inferred through GPS or other methods mentioned above, providing the absolute position on a map.

BLE avoids the possibility of tracking and considers contact epidemiologically relevant only. Another advantage is that the data collected shall be stored in the device directly. Conversely, geolocation data can offer additional information about the context in which the relevant contact has taken place, providing clues about the potential propagation of the virus.

DCT apps have been developed in different countries, but governments have pursued different strategies of surveillance. Among the most relevant apps pioneered, it should be mentioned, *inter alia*, Trace Together in Singapore (which uses BLE data), WeChat and Alipay in China (which process location data on the travel of users) and Corona100 in South Korea. The latter crosses geolocation data with other personal and non-personal data.

In Europe, both the European Commission and EDPB have stated their position about methods of DCT.

The former issued a Recommendation on a common Union toolbox for the use of technology and data to combat and exit from the Covid-19 crisis, in which the Commission expressed its preference for collecting proximity data instead of location data on position and movements of an individual.

The same position has been expressed in the Communication Guidance on Apps supporting the fight against Covid-19 pandemic in relation to data protection, in which there was expressed a preference for contact tracing utilising BLE, rather than the use of geolocation data.

The EDPB - in its Guideline on the use of location data and contract tracing tools in the context of the Covid-19 outbreak - expressed its concerns regarding DCT apps. The use of such instrument, according to EDPB, is therefore subject to the adoption of the following criteria: (a) voluntary usage, (b) DPIA before their development, (c) predilection for proximity data, (d) disclosure of information on who the infected has been in close contact with, (e) data minimisation and data protection by design, (f) encrypted identifiers generated by BLE and (g) anonymity of third users involved.

In the point of privacy protection, processing geolocation data continuously proves to be invasive.

Large-scale monitoring of location data paves the way to considerations about the balance between health protection and privacy. In those cases, it takes proportionality of means to pursue objectives. Taking into due account the peculiarities of location data, difficulties for their anonymisation and their revealing nature, the vast majority of privacy-oriented decisions lean towards the collecting of proximity data. But, from a comparative perspective, there are examples of geolocation apps that shall not have an intrusive effect on the life of individuals. For instance, the case of the app of Rhode Island comes to mind. Such an app provides a section named "my location diary", which allows the recalling of geolocation data that are locally recorded within twenty days, collected by other apps. Should the swab prove to be positive, users may decide to share, discretionally, such data with health authorities.

In conclusion, several difficulties stand between the collection of location data and their secure processing. Part of this is due to their revealing nature about sensitive information on the person involved and the difficulties encountered in their anonymisation process. The latter still leaves room for the reverse procedure, which allows data subjects to be identified.

However, the benefits of their usage cannot be denied.

The processing of those data thus needs to be calibrated to take the balancing between purposes pursued and personal data protection into account. Hence, in addition to all the necessary measures mentioned above, the theme of proportionality remains central. Until it is possible to reach a result consistent with the aim, an approach that does not jeopardise the data subject is desirable. Conversely, if the achievement of this aim prevails in the scale of fundamental principles, as the right to health compared to the right to privacy does - and there are no alternatives but to compress the latter - then the measures that will be adopted should be compatible with such scale of values.

Author Contribution Alberto Maria Gambino sects. 1 and 3.3, Davide Tuzzolino sects. 2, 3, 3.1and 3.2

References

Article 29 Working Party (2001) Opinion 8/2001 on the processing of personal data in the employment context. https://ec.europa.eu/justice/article-29/documentation/opinion-recommendation/files/2001/wp48_en.pdf

Article 29 Working Party (2009) Opinion 2/2009 on the protection of children's personal data. https://www.garanteprivacy.it/documents/10160/10704/1619292.pdf/1ab4d295-c2b9-405f-a2df-1e0f7ec9cfd1?version=1.0

Article 29 Working Party (2011) Opinion 13/2011 on Geolocation services on smart mobile devices. https://www.apda.ad/sites/default/files/2018-10/wp185_en.pdf

Baldassarre C (2020) La app Immuni al banco di prova, tra rispetto della privacy e difesa della salute pubblica. Danno e responsabilità 6:687–697

Bu-Pasha S et al (2016) EU law perspectives on location data privacy in smartphones and informed consent for transparency. Eur Data Prot Law Rev 2:312–323

Camardi C, Tabarrini C (2020) Contact tracing "Ordinario" e "straordinario" nella disciplina del diritto al controllo dei dati personal. Suppl Nuova Giurisprudenza Civile Commentata 3:32–39

Chambers Global Practice Guide (2020) Data protection and privacy

Cuffaro V, D'Orazio R (2020) La protezione dei dati personali ai tempi dell'epidemia. Il Corriere giuridico 6:729–739

Ekong I et al (2020) COVID-19 mobile positioning data contact tracing and patient privacy regulations: exploratory search of global response strategies and the use of digital tools in Nigeria. JMIR Mhealth Uhealth 8(4): https://doi.org/10.2196/19139

European Commission (2020a) Communication. Guidance on Apps supporting the fight against COVID 19 pandemic in relation to data protection. https://ec.europa.eu/info/sites/info/files/5_en_act_part1_v3.pdf

European Commission (2020b) Recommendation (EU) 2020/518 of 8 April 2020 on a common Union toolbox for the use of technology and data to combat and exit from the COVID-19 crisis, in particular concerning mobile applications and the use of anonymised mobility data. https://eur-lex.europa.eu/legal-content/EN/TXT/?uri=CELEX%3A32020H0518

European Data Protection Board (2020a) Guidelines 01/2020 on processing personal data in the context of connected vehicles and mobility related applications. https://edpb.europa.eu/our-work-tools/our-documents/guidelines/guidelines-012020-processing-personal-data-context-connected_it

European Data Protection Board (2020b) Guidelines 04/2020 on the use of location data and contact tracing tools in the context of the COVID-19 outbreak. https://edpb.europa.eu/sites/edpb/files/files/file1/edpb_guidelines_20200420_contact_tracing_covid_with_annex_en.pdf

European Union Agency for Network and Information Security (2017) Privacy and data protection in the mobile applications. A study on the app development ecosystem and the technical implementation of GDPR. https://pure.uva.nl/ws/files/42887337/22302384.pdf

Gray S (2020) A closer look at location data: privacy and pandemics. https://fpf.org/blog/a-closer-look-at-location-data-privacy-and-pandemics/

Istituto Superiore di Sanità - Bioethics COVID-19 Working Group (2020) Supporto digitale al tracciamento dei contatti (contact tracing) in pandemia: considerazioni di etica e di governance. https://www.iss.it/documents/20126/0/Rapporto+ISS+COVID-19+59_2020.pdf/c8611778-e4d8-2ec2-94e4-72c9107f84a2?t=1600695788673

Kędzior M (2021) The right to data protection and the COVID-19 pandemic: the European approach. ERA Forum 21:533–543

Organisation for Economic Co-operation and Development (2020) Tracking and tracing COVID: protecting privacy and data while using apps and biometrics. http://www.oecd.org/coronavirus/policy-responses/tracking-and-tracing-covid-protecting-privacy-and-data-while-using-apps-and-biometrics-8f394636/

Resta G (2020) La protezione dei dati personali nel diritto dell'emergenza Covid-19. Giustiziacivile.com. http://giustiziacivile.com/soggetti-e-nuove-tecnologie/editoriali/la-protezione-dei-dati-personali-nel-diritto-dellemergenza

Scantaburlo T et al (2020) Covid-19 and contact tracing apps: review under the European legal framework. arXiv:2004.14665

Ventrella E (2020) Privacy in emergency circumstances: data protection and the COVID-19 pandemic. ERA Forum 21:379–393

Rise and Fall of Tracing Apps

Giorgio Resta and Vincenzo Zeno-Zencovich

1 Introduction

When in March 2020, the COVID-19 pandemic erupted in Europe with thousands of casualties every day and extreme difficulty in preventing it from spreading even more violently, great hope was put in so-called 'tracing (or tracking) apps', already experimented in some Asian countries (China, South Korea, Singapore).

The idea behind the app (and in particular the Bluetooth Low Energy app) was simple: one downloaded the app in one's smartphone. If one resulted positive to the COVID-19 virus, one alerted the system, and all those who had downloaded the app and had been for a certain time (approx 15 min) and at a short distance (approx 1–2 m) from the infected person were warned of a possible occasion of contagion and invited to verify their positivity/negativity and to self-quarantine.

Rapidly the various European countries selected a model of the app. The EU issued lengthy normative texts meant to provide general guidance and promote uniformity and compliance with community laws.

When, towards the end of April 2020, the various applications started to roll-out, the response was extremely slow and limited. The number of those joining the programme increased, but certainly not skyrocketed even following the upsurge of the pandemic in its second and deadly wave in Autumn 2020. One can, very frankly, say that anti-COVID tracing apps have been an almost complete failure.

In this short paper, we would like to succinctly point out some of the possible causes and what lessons may be learnt for the future.

G. Resta · V. Zeno-Zencovich (✉)
Roma Tre University, Rome, Italy
e-mail: vincenzo.zenozencovich@uniroma3.it

G. Resta
e-mail: giorgio.resta@uniroma3.it

R. Senigaglia et al. (eds.), *Privacy and Data Protection in Software Services*,
Services and Business Process Reengineering,
https://doi.org/10.1007/978-981-16-3049-1_13

2 The Complexity of Legal Transplants

Since the very beginning of the European debate on tracing applications, the East-Asian experience has been taken as a reference model in a quite simplistic manner, without carefully reflecting on the peculiar institutional framework surrounding the use of tracing apps in those systems and conditioning the efficacy of a legal transplant. First of all, it deserves to be noted that tracing apps were not the only and not even the most important of the digital solutions adopted in the East with the aim of countering the pandemic. Migration maps created by integrating different sources of data, last generation screening technologies (such as body temperature scans), AI models applied to health data for the purposes of diagnosis and risk prediction, electronic monitoring of home-quarantined individuals, health QR codes, virtual care platforms, robots for personal care in hospitals, all of the above are just some examples of the panoply of digital technologies effectively deployed since the very first stage of the pandemic. Secondly, to correctly appreciate the effectiveness of East-Asian strategies, one should keep in mind that COVID-19 was just the last episode of a long wave of health crises triggered by contagious diseases experienced in recent times in that region. China, Hong Kong, South Korea were already faced with the need to restructure the whole framework of disease control first in 2002 following the outbreak of the SARS epidemic, and later in 2013 of the MERS (which hit South Korea). This led to a revision and modernization of the respective legislations on disease control, which proved extremely helpful for the fight against COVID-19. In China, the general *Law on Prevention and Treatment of Infectious Diseases* (1989) was deeply revised in 2004 and later amended in 2013. Together with the 2003 *Law on Emergency Response* and the 2003 regulation on *Contingent Public Health Emergencies*, this statute provided the main legal framework for rapidly adopting a set of wide-ranging measures, such as the blockading of entire areas, cities or regions of the country, the building of dedicated hospitals and the development of online healthcare platforms (measures rapidly put in place in the cities most hardly hit by the pandemic, such as Wuhan). In South Korea, the *Act on Infectious Diseases Prevention and Control* was amended in 2015, with the aim of improving the responses to the possible outbreak of new contagious diseases. The new Articles 34 *bis* and 76 *bis* of the Act make massive recourse to various sources of data for tracking purposes possible. In particular, they grant public authorities the power to access a wide gamut of personal data—geolocation data, communications metadata, history of purchases and financial data, health data, video surveillance footage—with the aim of tracking the patterns of the disease and informing the public through detailed migration maps. As a result of this background, South Korea and China were extremely fast and efficient in adopting a wide range of measures—physical and digital—as soon as the COVID-19 epidemic erupted. Lastly, it is worth underlining that tracing applications are embedded in a cultural and legal framework whose features mark a stark contrast, from the several points of view, with the European tradition. Among such features are the following: (a) quasi-compulsory use of digital tools, as evidenced by the Chinese resorting to the QR Code as a requirement to access public places; (b) loose application of data protection principles vis-à-vis public authorities, as evidenced by the Chinese and Singaporean experience; (c) adoption of centralized models of tracing applications and access by health authorities to proximity data (Singapore); (d) strict surveillance and harsh enforcement of quarantine obligations. Since most of these features appear to be in contrast not only with the GDPR, as will be later detailed, but also with the general framework of European fundamental rights (see in particular art. 8 ECHR and art. 8 European Charter Fundamental Rights), the transplant of the techno-legal model 'tracing app' was necessarily a selective and highly ineffective one.

3 Technical Inadequacies

It would appear that one of the essential features of all tracing apps was that the smartphone in relation to proximity should have activated the Bluetooth application enabling a reciprocal connection. However, Bluetooth is an energy-consuming technology which renders it not very attractive especially for those who are in possession of old devices and are in open spaces. Furthermore, Bluetooth compared to the GPS has the advantage of being a less privacy-intrusive technology, as it does not disclose the location of its user, but only the distance and duration of the exposure. However, tracing applications based on Bluetooth—a technology developed to make communications between two devices possible—have strong limitations in terms of precision of measurements (more so if the sensors built in the smartphones are not of the last generation) and are prone to false positives. For instance, the presence of a wall or a Plexiglas shield between the two devices would not be recorded by the system. Lastly, BLE applications can detect proximity as long as the smartphone is switched on and carried by its user; if these conditions are not given, then the tracing app would be unable to detect proximity, both active and passive. It is no wonder, therefore, that alternative solutions, such as cheap wearable devices independent from any smartphone, have been proposed and carefully considered by the decision-makers.

4 Digital Divide

The tracing apps were developed for the current generation of smartphones. This clearly cut out all the holders of less recent devices, and in particular traditional mobile phones, not connected to the internet, and very common among elderly persons, the most exposed to infection and its dire consequences. Obviously, it also cut out all the persons not owning a cell phone, like children and the very poor.

5 Organizational Failures

To be effective, a tracing app requires around it—as anticipated above—an efficient health prevention system. Not only is it necessary that the person who has resulted positive in the COVID-19 test alerts the system, but it is necessary that this rapidly detects those who have been in his/her proximity. If the alert arrives many days later, the prevention effect is substantially watered down, with a chain reaction: those who have been in contact with a person who results positive will know if they have been infected only several days later, and in the meantime, may have infected many others. The Italian experience is illustrative. At least during the first and the second wave of the pandemic, the Italian screening system was under pressure, and due to serious organizational deficiencies, people had to wait long hours to get tested and several days to get the results. This meant that an alert could have been sent days after the appearance of symptoms. Furthermore, doctors and other personnel in charge of the unlocking of the app and the sending of an alert had not been properly instructed and, in several cases, proved unable to initiate the notification proceeding.

6 The GDPR Totem

However, the principal reason for the failure of tracing apps in the EU appears to be the concerns—real or supposed—related to respect of privacy, and more specifically compliance with the GDPR. Clearly tracing apps collect, directly or indirectly, personal data on the location, movements and activity of those who have downloaded them. If and when an individual test is positive and alerts the system, the data are of sensitive nature and therefore, undergo special requirements in its processing. The consequences that the GDPR has had on the roll-out of the tracing apps appear to be manyfold:

(a) The GDPR is a mastodontic piece of legislation that covers many layers of public and private actions. The COVID-19 pandemic has shown that it is overly constraining and to a certain extent unworkable in an emergency. With the risk of enormous sanctions (economic, administrative and even criminal), an ordinarily cautious public or private decision-maker is naturally nudged towards a work-by-the-rule approach. The very aggressive stance that EU authorities and national data protection agencies have had during these last years has had a highly deterrent—if not chilling—effect. Just to give one example, the processing of data for scientific research, and in particular, the cross-border transfer of samples and data for studies related to COVID-19, has been obstructed and made more burdensome given the not-so-clear legal basis offered by Article 49 GDPR.

(b) The clearest result of such a privacy-above-all approach is the decision of the producers of the main anti-COVID-19 tracing app, Google and Apple, to favour a decentralized model: the data concerning the holder of the smartphone and his or her contacts remain strictly under the holder's control. Only if he or she decides to inform on a COVID-19 positivity, does the system enter in action sending the alerts to the eventual contacts. Furthermore, the collection of GPS data from telecommunication providers has been limited to exceptional cases, and this has impeded the creation and disclosure of migration maps such as those employed in South Korea. This solution is highly inefficient because it relies on individual choices for the success of a public health strategy. However, it is understandable that the companies—constantly under fire as members of the GAFAM 'villain group'—were not willing to develop different apps that hypothetically could have brought them into further regulatory troubles. Also, they had a specific interest in entering the promising commercial field of eHealth and at the same time marketing their position as inflexible guardians of privacy, who would never agree to transfer sensitive data to the governments.

(c) Data protection is presented as a distinctive feature of the EU constitutional (Article 7 of the CFREU) and institutional (EDPS, EDPB) framework. It is one of the bastions of the 'European fortress' used to protect the EU from data appropriation and exploitation by big-tech giants from the East (China) and the West (the already mentioned GAFAM). With all the rhetoric behind personal data protection, enhanced by scores of hardline decisions of the CJEU, a flexible and realistic approach was untenable.

(d) The rhetoric of the GDPR has invaded and infected public and private discourse: millions of citizens who every second of their daily life provide thousands of personal data to all the private online business (not only the big-tech giants but also any provider of apps or online services), enabling them to profile them in every aspect, raised the alarm over a revised and incumbent version of the Orwellian Big Brother. The individual right to privacy became the antagonist of public health concerns which had to surrender. The cleavage between individualist (and selfish) EU societies and community-oriented (and rigidly governed) Eastern ones has become ever wider.

(e) The result is that an app that in order to be effective needed to be downloaded and kept constantly in function by at least two-thirds of the adult population, in its peak reached not more than 30% of the citizens of a well-organized and disciplined country such as Germany and an average 15% in the others. It should have been compulsory and centralized (along the lines of the East-Asian experience). Left to self-determination, it was doomed to fail.

7 The Issue of Public Trust

The overall management of the COVID-19 pandemic has raised deep concerns on the ability of governments to face and counter such an unprecedented emergency. At the end of the day, the only substantive remedies were those of the past centuries against the plague and deadly fevers: lockdowns and quarantine. The transfer of all substantive powers to the central government, the ancillary role of Parliaments, widespread deference of the judiciary towards the decision of the public authorities, a substantial suspension of constitutional rights (in particular of circulation and assembly), have determined a significant and widespread mistrust by citizens towards decision-makers. In a democratic system, the promotion of public goals passes necessarily through widespread compliance with the rules introduced. Tracing apps were outside this picture and were seen as only a further burden on an already significantly impaired way-of-life. Furthermore, it is worth reflecting on the fact, registered by many pollsters, that people had shown more confidence in the idea that proximity data were collected by Google and Apple—pursuant to the decentralized model—than by the governments, under the original centralized model. This is not only telling of the distrust of governments, but also of the disregard for the strong commercial reasons behind the move of Google and Apple, which would be silly to regard as a purely altruistic act.

8 Some Lessons for the Future

(a) The GDPR obstructs—formally and substantially—most policies, whether digital or organizational, to contrast this and future pandemics and to unlock the potential of scientific research. Either public health is put in the same special regime that already covers police and criminal investigations (see Directives 680 and 681/2016) or it will risk succumbing to the 'fundamental' right to individual privacy.

(b) These concerns are quite obvious when one looks at a much more simplified issue such as that of 'vaccination passports' which would, finally, give back to European citizens one of their most cherished freedoms, that of unrestricted movement between the 27 Member States. It is no wonder that as soon as the first proposals were disclosed to the public, national data protection authorities immediately raised strong objections based on the GDPR.

(c) Coordination of health policies is still at a primitive stage in the EU. All the obstacles one has experienced over the last year in coordinating responses highlight the urgent need for pragmatic output policies. The economic and social cost that the whole of Europe is paying for a piecemeal approach, risk to broaden disaffection towards the EU.

(d) In this perspective, eHealth services (surely among the best and trustworthy in the world) should become one of the main short-term goals, together with economic recovery.

(e) Finally, the rise and fall of tracing apps should make us wary of the daily prophecies of the Big Data superpowers. Never, in the past, has so much information been collected throughout the world, from millions and millions of cases, concerning an illness. However, at the end of the day, the battle will be won not because some algorithm has provided us with a predictive analytic solution, but by brick-and-mortar medical research, based on trial-and-error, statistical samples of the population, laboratory and on-the-field experimentation. Fighting a pandemic is nowhere near guessing the 'sentiment' of social media goers and profiling consumer habits.

References

Ayres I, Romano A, Sotis C (2020) How to make COVID-19 contact tracing apps work: insights from behavioral economics. Available at SSRN https://papers.ssrn.com/sol3/papers.cfm?abstract_id=3689805

Bonsall D, Parker M et al (2020) Sustainable containment of COVID-19 using smartphones in China: scientific and ethical underpinnings for implementation of similar approaches in other settings. https://github.com/BDI-pathogens/covid-19_instant_tracing/blob/master/Policy%20forum%20-%20COVID-19%20containment%20by%20herd%20protection.pdf

Cho H, Ippolito S et al (2020) Contact tracing mobile apps for COVID-19: privacy considerations and related trade-offs. https://arxiv.org/abs/2003.11511

Della Morte G (2020) Quanto Immuni? Luci, ombre e penombre dell'app selezionata dal Governo italiano. Diritti umani dir int 14(2):303–336

Du L, Wang M (2020) Chinese CoViD-19 epidemic prevention and control measures: a brief review. Biolaw J. https://doi.org/10.15168/2284-4503-20201S

European Commission (2020a) Guidance on Apps supporting the fight against COVID 19 pandemic in relation to data protection. https://eur-lex.europa.eu/legal-content/EN/TXT/?uri=CELEX%3A52020XC0417%2808%29

European Commission (2020b) Recommendation on a common Union toolbox for the use of technology and data to combat and exit from the COVID-19 crisis, in particular concerning mobile applications and the use of anonymised mobility data. https://op.europa.eu/it/publication-detail/-/publication/1e8b1520-7e0c-11ea-aea8-01aa75ed71a1/language-en

European Commission (2021a) Proposal for a EU Regulation on a framework for the issuance, verification and acceptance of interoperable certificates on vaccination, testing and recovery to facilitate free movement during the COVID-19 pandemic (Digital Green Certificate). https://ec.europa.eu/commission/presscorner/detail/en/ip_21_1181

European Commission (2021b) Proposal for a EU Regulation on a framework for the issuance, verification and acceptance of interoperable certificates on vaccination, testing and recovery to third-country nationals legally staying or legally residing in the territories of Member States during the COVID-19 pandemic (Digital Green Certificate). https://www.europeansources.info/record/proposal-for-a-regulation-on-a-framework-for-the-issuance-verification-and-acceptance-of-interoperable-certificates-on-vaccination-testing-and-recovery-to-third-country-nationals-legally-staying-or

European Data Protection Board (2020a) Guidance 4/2020 on the use of location data and contact tracing tools in the context of the COVID-19 outbreak. https://edpb.europa.eu/sites/edpb/files/files/file1/edpb_guidelines_20200420_contact_tracing_covid_with_annex_en.pdf

European Data Protection Board (2020b) Guidance n. 3/2020 on the processing of health data for the purpose of scientific research in the context of the Covid-19 outbreak. https://edpb.europa.eu/sites/edpb/files/files/file1/edpb_guidelines_202003_healthdatascientificresearchcovid19_en.pdf

European Parliament (2020) Resolution of 17 April 2020 on EU coordinated action to combat the COVID-19 pandemic and its consequences. https://www.europarl.europa.eu/doceo/document/TA-9-2020-0054_EN.html

Ferretti L, Wymant C et al (2020) Quantifying SARS-CoV-2 transmission suggests epidemic control with digital contact tracing. Science 368(6491):eabb6936

Findlay M, Remolina N (2020) Regulating personal data usage in Covid-19 control conditions. SMU Centre for AI & Data Governance Research Paper No. 2020/04

Garante per la protezione dei dati personali (2021) Press Communiqué 1-3-2021 (doc. web 9550331), No a 'pass vaccinali' per accedere a locali o fruire di servizi senza una legge nazionale. https://www.garanteprivacy.it/web/guest/home/docweb/-/docweb-display/docweb/9550331

Geddie J, Aravindan A (2020) Singapore plans wearable virus-tracing device for all. Reuters. https://www.reuters.com/article/us-health-coronavirus-singapore-tech/singapore-plans-wearable-virus-tracing-device-for-all-idUSKBN23C0FO

Greenleaf G, Kemp K (2020) Australia's 'COVIDSafe App': an experiment in surveillance, trust and law. University of New South Wales Law Research Series 999. Available at SSRN https://papers.ssrn.com/sol3/Delivery.cfm/SSRN_ID3589317_code57970.pdf?abstractid=3589317&mirid=1

Greenleaf G, Kemp K (2020) Australia's COVIDSafe experiment, phase III: legislation for trust in contact tracing. UNSW Law Research. Available at SSRN https://papers.ssrn.com/sol3/papers.cfm?abstract_id=3601730

Gyooho L (2020) Legislative and administrative responses to COVID-19 virus in the Republic of Corea. Available at SSRN https://papers.ssrn.com/sol3/Delivery.cfm/SSRN_ID3587595_code1390660.pdf?abstractid=3587595&mirid=1https://dx.doi.org/10.2139/ssrn.3587595

Hipgrave D (2011) Communicable disease control in China: from Mao to now. J Glob Health 1(2):224–238

Holmes A (2020) Singapore is using a high-tech surveillance app to track the coronavirus, keeping schools and businesses open. Here's how it works. Business Insider, 24 Mar 2020

Istituto Superiore di Sanità (2020) Rapporto ISS-Covid 19 n. 59/2020, Supporto digitale al tracciamento dei contatti (contact tracing) in pandemia: considerazioni di etica e di governance. https://www.iss.it/rapporti-covid-19/-/asset_publisher/btw1J82wtYzH/content/rapporto-iss-covid-19-v.-59-2020-supporto-digitale-al-tracciamento-dei-contatti-contact-tracing-in-pandemia-considerazioni-di-etica-e-di-governance.-versione-del-17-settembre-2020

Kritikos M (2020) Ten technologies to fight coronavirus. EPRS, Brussels

Kuner C (2020) Data crossing borders: data sharing and protection in times of Coronavirus, VerfBlog, 15 Apr 2020

Liu W et al (2020) Response to the COVID-19 epidemic: the Chinese experience and implications for other countries. Int J Environ Res Public Health 17:2304

Renda A, Castro R (2020) Towards stronger EU governance of health threats after the Covid-19 pandemic. Eur J Risk Regul 11(2):273–282

Resta G, Data and Territory. The impact of the "local" in the regulation of digital technologies and algorithmic decision-making, in Essays in Honour of Mads Andenas, forthcoming

Savona M (2020) The saga of the Covid-19 tracing apps: what lessons for data governance?. SPRU working paper series, n. 10/2020. Available at SSRN https://papers.ssrn.com/sol3/papers.cfm?abstract_id=3645073

Sharon T (2020) Blind-sided by privacy? Digital contact tracing, the Google/Apple API and big tech's newfound role as global health policy makers. Ethics Inf Technol 18:1–13. https://doi.org/10.1007/s10676-020-09547-x

Tian H et al (2020) An investigation of transmission control measures during the first 50 days of the Covid-19 epidemic in China. Science 368(6491):638–642

Wang Z (2017) Systematic government access to private-sector data in China. In: Cate F, Dempsey J (eds) Bulk collection. Oxford University Press, Oxford

Wendehorst C (2020) Covid-19 apps and data protection. In: Hondius E et al (eds) Coronavirus and the law in Europe. Intersentia. https://www.comparativecovidlaw.it/2020/09/02/covid-19-apps-and-data-protection

Whitelaw S et al (2020) Applications of digital technology in Covid-19 pandemic planning and response. Lancet Digit Health 2(8):435–440

World Health Organisation (2020) Contact tracing in the context of COVID-19 (Interim Guidance). https://apps.who.int/iris/handle/10665/332049

Xu T et al (2020) China's practice to prevent and control COVID-19 in the context of large population movement. Infect Dis Poverty 9(115):1–14

Zastrow M (2020) Coronavirus contact-tracing apps: can they slow the spread of COVID-19?. Nature https://doi.org/10.1038/d41586-020-01514-2

Zhang L (2020) Measures to control infectious diseases under Chinese law. https://blogs.loc.gov/law/2020/01/falqs-measures-to-control-infectious-diseases-under-chinese-law

Zhao Q et al (2020) On the accuracy of measured proximity of bluetooth-based contact tracing apps. In: Park N et al (eds) Security and privacy in communication networks. Springer, Cham, p 49

Privacy, Software and Insurance

Sara Landini

1 Definitions

What is software and when we can use properly the term software applied to insurance? Software comprises the entire set of programs, procedures, and routines associated with the operation of a computer system. The term differentiates these instructions from hardware, the physical components of a computer system.

Software is often divided into categories. System software is a computer program designed to run a computer's hardware and application programs and coordinates the activities and functions of the hardware and software. In addition, it controls the operations of the computer hardware and provides an environment or platform for all the other types of software to work in.

Application software is a computer software package that performs a specific function for an end-user or, in some instances, for another application. An application can be self-contained or a group of programs. The program is a set of operations that runs the application for the user. Applications use the computer's operating system and other supporting programs, typically system software, to function. Application software is different from other software that might come pre-bundled with a computer's operating system, such as a utility.

Application software refers to user-downloaded programs that fulfil a want or need. They include office suites, database programs, web browsers, word processors, software development tools, image editors, and communication platforms. Another category of software is the set of utilities, which are small, useful programs with limited capabilities. Additionally, some utilities come with operating systems. Like applications, utilities tend to be separately installable and capable of being used independently from the rest of the operating system.

S. Landini (✉)
University of Florence, Florence, Italy
e-mail: sara.landini@unifi.it

R. Senigaglia et al. (eds.), *Privacy and Data Protection in Software Services*,
Services and Business Process Reengineering,
https://doi.org/10.1007/978-981-16-3049-1_14

System software includes operating systems and any program that supports application software. It is also important to distinguish the term software from software engineering. Connected to this is the idea that although the terms 'computer science' and 'software engineering' are often used interchangeably, they are not the same. Computer science is the field of computing that deals with the study, implementation, and analysis of algorithms. Software engineering, on the other hand, focuses on applying structured engineering principles to the development of software. In any case, software engineering is directly related to computer science, where engineers take systematic and disciplined methods to the development, operation, and maintenance of software.

We will focus on application software used for the specific needs of the insurance industry.

Now, what is an insurance contract, and why is software of interest in the insurance industry? An insurance contract is an agreement between an insurance company and the insured under which one party (the insurer), in consideration of receipt of a premium, undertakes to pay money to another person (the insured) on the happening of a specified event (as, for example, in death or accident or loss or damage to property). Central to any insurance contract is the insuring agreement, which specifies the risks that are covered, the limits of the policy, and the term of the policy (Clarke 2009). Additionally, all insurance contracts specify conditions, which are requirements of the insured, such as respecting a special conduct code or installing a software; limitations, which specify the limits of the policy, such as the maximum amount that the insurance company will pay; and exclusions, which specify what is not covered by the contract.

Insurance contracts have an additional requirement that they are in legal form. Insurance contracts are regulated by state law, so insurance contracts must comply with these requirements. The state may stipulate that only certain forms may be used for certain types of insurance or that the contract must have certain provisions.

If a contract lacks any of these essential elements, then the contract is deemed null and void and will not be enforced by any court.

In insurance, an offer is typically initiated by the insurance applicant through the services of an insurance agent, who must have the authority to represent the insurance company, by filling out an insurance application. However, the insurance application sometimes can be filed directly with the insurance company through its website or indirectly through a broker. A broker is a person or firm who arranges transactions between an insurer and a client for a commission when the deal is executed. Neither role should be confused with that of an agent—one who acts on behalf of a principal party in a deal. How the offer is accepted depends on whether the insurance is for property, liability, or life insurance. These are the types of insurance contracts.

With regard to insurance distribution, we have to recall that on 20 January 2016, the European Parliament and the Council of the European Union issued Directive (EU) 2016/97, the Insurance Distribution Directive (IDD).

The IDD introduced new duties on distributors and reinforced some past duties:

- Expanding the scope from agents and brokers by adding all sellers of insurance products, including insurance manufacturers that sell directly to customers and market participants who sell insurance on an ancillary basis (subject to the proportionality conditions).
- Having stricter requirements surrounding conflicts of interest and remuneration disclosures.
- Making special disclosure requirements for bundled products and other product oversight requirements similar to those of MiFID II, Directive (EU) 65/2014, i.e., the legislative framework instituted by the European Union to regulate financial markets in the bloc and improve protections for investors.
- Giving additional requirements for insurance-based investment products (IBIPs) and the introduction of an Insurance Product Information Document (IPID) for non-life insurance products.
- Adding new provisions regarding cross-border activity (freedom to provide services and freedom of establishment).
- Having stricter administrative sanctions and other measures, including pecuniary sanctions.

The IDD also introduced product oversight and governance requirements similar to MiFID II for all insurance products. The approval process for each insurance product should be defined as proportionate to the nature and function of the insurance products that are about to be sold to customers. The process should incorporate the identification of the target market, the risk assessment, and also assure that the distribution strategy is aligned with the identified market. Regular reviews are expected to check that products remain effectively distributed and consistent with the objective of the respective target markets. There are exemptions for insurance of large risks.

As we have seen, insurance production and distribution are characterised by moments of risk assessment, which can be facilitated by the presence of software, and by a strong process and compliance with formal rules that can be facilitated by software (Marano and Noussia 2020).

2 Software in Distribution: Profiling Clients, Checklists, and Compliance

The insurance industry is charged with protecting and supporting its customers in their most challenging times. Any breach of insurance industry regulation is compounded not only by regulatory consequences (administrative sanctions) but also by the damage inflicted onto the individual customer or corporate client. Regulatory compliance and risk management for insurance companies require organisations to abide by a comprehensive 'know your customers rule' standard, impeccable privacy, and anti-money laundering and anti-corruption practices.

Transparency is key to all businesses but especially insurance providers. Strict adherence to insurance industry compliance needs to be woven into daily business practices. Operating above board on issues like data security and privacy, case and complaint management, and all forms of fraud management are essential to building and keeping customers as well as aligning with insurance industry regulatory compliance.

Insurance compliance software enables insurance companies and insurance intermediaries to meet compliance regulations efficiently and effectively. They use these solutions to reduce non-compliance events, establish effective compliance processes, and maintain strict, auditable records for compliance officers.

Insurance compliance solutions typically contain policy and procedure management (Marano and Noussia 2020), tools to manage compliance policies and procedures, insurance-specific regulatory intelligence capabilities, incident management, complaint management, task management, audit trails for compliance officers, workflow management, reporting, and regulatory intelligence features. These provide a comprehensive set of tools for insurance intermediaries that are used to govern all of their compliance-related tasks.

And like any other large business, insurers face all the usual requirements to protect personal information under rules such as the Regulation (EU) 2016/679 (GDPR) and State consumer protection laws. In all these cases, the application of software deals with clients' personal data processing (Forgò et al. 2017).

As said, it is important to meet customer's needs, and, in insurance contracts, it is difficult to assess risk and customers' needs (capability of the customers to retain the risk on their own).

Generally speaking, there are different reasons why a company opts for a customer profiling tool, but the main reason is so that companies can focus their sales and marketing efforts on generating high-quality sales leads(Helberger 2016; McGurk 2019) .

This is why creating customer profiles is so important. A customer profiling tool is the means to create a portrait of customers to help companies make design decisions concerning their service.

Nowadays there is a comprehensive range of good data profiling software solutions (even free for download).

Data profiling is an assessment of data values within a given data set for uniqueness, consistency, and logic—the key data quality metrics (Hoeren and Kolany-Raiser 2018).

Data profiling is the first step of a data quality assessment that identifies business rules violations and anomalies. It involves activities of analysing one's data contents and structure.

Data profiling software and techniques provide companies with the ability to analyse large amounts of data quickly, in no time.

Additionally, Big data can enable major changes in the way claims are handled. If claims handlers have access to the data and can use it in a meaningful way, they are able to paint a much clearer picture and investigate more accurately. Drawing on the example of the black box in cars, insurers can see where and when the accident occurred and what speed the insured was moving, therefore giving them a much clearer indication of the validity of the claim (Nazzaro and Landini 2020; Maurer et al. 2017).

3 Software to Reduce Risk and Monitor Claims

Software can also be used to help the insured to reduce the risk, and the insurer can introduce into the general conditions of the contract clauses on the mandatory installation/application of software to reduce the risk.

With regard to cyber-risk insurance coverages (business interruption due to cyber-attack, liability for damage to customers' data due to cyberattack), this solution can reduce the risk.

Cybersecurity remains a challenge for businesses across industries. Cyberattacks such as malware, ransomware, and phishing can breach enterprise systems and networks to steal confidential client and business data. Also, cybercriminals are continuously coming up with new attacking tools and techniques, making cybersecurity the need of the hour for all businesses.

Cybersecurity software can help protect computer systems, IT networks, software platforms, and mobile applications from hacking attempts. It uses security technologies such as encryption, endpoint protection, and multi-factor authentication to protect one's enterprise data in real-time from cyberattacks.

A wide range of cybersecurity software tools is available on the market. Cybersecurity software is a software solution that identifies vulnerabilities and potential threats to protect business systems, applications, and networks from cyber threats, including viruses, ransomware, and phishing attempts. It uses a combination of technologies such as firewall protection, data encryption and backup, incident response, and website scanning to prevent unauthorised access and ensure real-time enterprise security.

There are many types of cybersecurity software solutions: data encryption tools, web vulnerability scanning tools, network defence tools, penetration testing tools, antivirus software, and firewall software. Application security, information security, network security, operational security, and disaster recovery are some common business applications of these tools.

There are different features offered by cybersecurity software solutions:

- Vulnerability scanning examines systems, software, and networks at regular intervals to detect and report on any new or existing security vulnerabilities, such as viruses and malware.
- Threat mitigation employs security techniques to detect existing threats, reduce the impact of the detected threats, and prevent the occurrence of new threats. All identified security threats are quarantined to prevent contamination of other files and data.

- Incident management sets up a plan of action to follow in case a security incident is identified, logs incidents by priority and diagnoses the issue to reduce downtime.
- Data encryption cyphers business data, so it can be accessed or decrypted only by users that have the encryption key (i.e., a password or passcode).
- Single sign-on uses a single set of login credentials (e.g., a username and password) to access multiple software applications or platforms.
- Two-factor authentication sets up a dual authentication mechanism to allow users access to business data and applications. All users have to verify their identity using two sets of credentials (e.g., mobile push authentication along with the standard username and password).

Among the benefits of cybersecurity tools in terms of cyber-risk reduction, we can recall the following:

- Protection of sensitive business data: cybersecurity software encrypts enterprise data to protect it from hacking attempts by unauthorised users. With encryption, data is converted into an unrecognizable, coded format, which can be unlocked only by users who have the encryption key (i.e., a password or passcode).
- Maintenance of secure computer networks: in the majority of cases, cyberattacks are launched through an organisation's computer network. Cybersecurity solutions identify malicious network activities and immediately send a notification, so appropriate action can be taken. They use various security techniques, such as vulnerability scanning, threat detection, and firewalls, to monitor your networks in real-time and prevent attackers from stealing sensitive data.

Regarding recent trends in the cybersecurity software market, it is important to underline the increasing use of artificial intelligence (AI) and machine learning (ML) to detect cyber attacks in real-time. Manual and semi-automated threat detection techniques are not able to keep up with today's constantly evolving cyberattack landscape. In such scenarios, AI and ML technologies are being used to bring the incident response time down to a few seconds via real-time threat intelligence and data security. AI and ML capabilities are directly deployed at network endpoints, such as mobile phones, laptops, desktops, tablets, and servers, to detect and combat threats in real-time.

It is also possible to use software installed on a car to monitor the driving habits of the driver (Glancy 2012). This is the case of 'Pay how you drive' (PAHD) insurance. PAHD insurance is a special type of motor vehicle insurance that takes how a person drives into consideration. This simply means that a person's driving habits dictate their premiums, i.e., from speeding, braking, parking, positioning, stops, etc. If the subjects happen to be rough drivers who speed, brake suddenly, and/or position themselves near obstacles/objects, they will obviously pay higher premiums compared to the careful driver who breaks gradually and leaves enough space between the car and objects. PAHD insurance is simply aimed at charging premiums according to individual driving habits. PAHD insurance considers all possible factors to ensure car owners are charged fair premiums.

There are many advantages of pay how you drive insurance:

1. It provides useful information: one of the main benefits of PAHD insurance is the information collected. Telematic tracking devices collect a lot of vital information, i.e., speed, car performance, concentration, braking, etc., which helps drivers evaluate their driving skills and take necessary measures;
2. It permits fair premiums: the fact that a driver pays a premium based on reasonable parameters, i.e., their driving skills and time, makes the amount of premiums charged justified. A person pays premiums according to fair experience;
3. It makes motor insurance cheaper for safer drivers;
4. It makes driving environmentally sustainable (Mahmood and Sayers 2015). It is possible, for instance, to use eco-driving as a model of risk discrimination in the case of motor insurance and on some related legal constraints. Several studies completed at an international level indicate a direct connection between efficient drivers and those drivers with fewer preventable accidents. The word eco-driving commonly indicates the combination of some driving techniques:
   (a) Maintenance. Key parameters to maintain are proper tire pressure, wheel alignment, and engine oil with low kinematic viscosity.
   (b) Driving lighter and/or lower-drag vehicles and minimising the number of people, cargo, tools, and equipment carried in the vehicle (removing common unnecessary accessories such as roof racks, brush guards, wind deflectors, etc., driving with the fuel tank mostly empty and tanking more frequently).
   (c) Maintaining an efficient speed. Optimal efficiency can be expected while cruising with no stops, at minimal throttle, and with the transmission in the highest gear.
   (d) Optimal choice of gear (in case of manual transmission).
   (e) Experts recommend accelerating quickly and smoothly.
   (f) A driver may further improve the economy by anticipating the movement of other traffic users. For example, a driver who stops quickly or turns without signalling reduces the options another driver has for maximising their performance.
   (g) Using air conditioning as required by the occupants and not continuously.

4 Key Points on GDPR and Insurance

All the above-mentioned applications of software in the insurance industry concern personal data processing (Cappiello 2018; Pype et al. 2017; Schultze and Staudenmayer 2016; Schwartz 1999).

The GDPR (General Data Protection Regulation), the European regulation on personal data protection, entered into force on 24 May 2016 and applies since 25 May 2018, is directly applicable in all Member States and covers insurance and insurers in two ways: business and compliance.

The GDPR is applicable to personal data processing carried out by a data controller or data processor based in the EU, as well as to personal data processing carried out by a data controller or data processor outside the EU, where such processing involves the supply of goods or services or the monitoring of the conduct of data subjects located in the EU. Therefore, insurance brokers or non-EU companies selling policies to EU citizens will be subject to the application of the GDPR as well.

As we have seen, insurance has always been based on data collection, which today is either intrinsically digital or dematerialised (i.e., digitalised). Personal data, often sensitive, is increasingly abundant thanks to new technologies for collection (such as smartphones or wearables, for example) and the need of structuring and making them leverage to keep up with today's market, which requires new products, slimmer and more personalised, on-demand policies, micro policies, etc.

GDPR concerns companies of all sizes. In fact, there are no exclusions by sector or corporate dimension from the applicability of the European Regulation apart from the processing register (Art. 30 of the GDPR). What matters is whether the company, as data controller or data processor, deals with personal data of data subjects located in the EU. Therefore, any requirement involving a large company is applicable to the emerging 'InsureTech' phenomenon.

With respect to past privacy protection systems, some novelties have been introduced by the GDPR. First of all, the approach to the regulation is that it is no longer prescribed, i.e., does not set what must (or must not) be done to be compliant, rather it defines specific targets to be achieved to guarantee the protection of personal data, on which the regulation has been construed through a series of steps and provisions driving the company through the adjustment process.

It is then important to determine who is the controller and who is the processor in the new GDPR terminology. The GDPR defines a controller as the natural or legal person, public authority, agency, or other body which, alone or jointly with others, determines the purposes and means of the processing of personal data. The GDPR defines a processor as a natural or legal person, public authority, agency, or other body which processes personal data on behalf of the controller.

Moreover, the GDPR introduced the concept of accountability of the data controller, which must be able to prove that the principles set out in Article 5 of the GDPR (lawfulness, correctness, and transparency in data processing; limitation of the purposes of processing; minimisation and accuracy of the data processed; integrity and confidentiality as well as limitation about data retention), applied to all relevant fulfilments and obligations, have been complied with. This approach is consistent with the introduction of the concept of 'Privacy by Design and by Default', pursuant to Art. 25 of the GDPR.

Additionally, the concept of 'Privacy by default' has been introduced, which means that data controllers must implement appropriate technical and organisational measures to ensure that only personal data necessary for a specific purpose will be processed.

Another novelty, and one of the biggest challenges that the insurance industry has to face in the new system, is the concept of data portability introduced under Article 20 of the GDPR. Data subjects will now have the right to receive any personal data concerning them, which they have previously provided or have observed, in a 'commonly used and machine-readable format' and have the right to transmit that data to another controller. This only applies to automatic processing and when personal data is being processed under the lawful basis of consent or performance of a contract.

Furthermore, a new consent form is another big task for insurers in the context of the GDPR. Special duties relating to the processing of personal data and particularly of 'sensitive data' (i.e., in the case of health insurance).

Problems in the case of joint-controlling of data can emerge. Article 26 of the GDPR introduces the concept of joint-controllers where there are two or more controllers that jointly determine the purposes and means of processing. As we have seen, insurers have relationships with numerous third parties, such as agents and brokers. In this case, insurers need to look at arrangements they have with third parties to determine if this is a controller-to-controller or controller-to-processor relationship.

Many new requirements have been introduced with regard to transparency, which is another key concept in the GDPR. One of the major challenges emerging in the insurance industry is the requirement under Article 14 to provide information where personal data has not been obtained from the data subject.

The GDPR poses also problems regarding the necessity to appoint a DPO (Data Protection Officer) to insurers and insurance intermediaries. DPO is responsible for overseeing a company's data protection strategy and its implementation to ensure compliance with GDPR requirements.

According to Article 37(1) of the GDPR, DPO must be appointed if:

- The relevant data processing activity is carried out by a public authority or body;
- The *core activities of the relevant business involve regular and systematic monitoring of individuals*, on a large scale;
- The core activities of the relevant business involve the processing of sensitive data, or data relating to criminal convictions and offences, on a large scale.

The statement "the core activities of the relevant business involve regular and systematic monitoring of individuals" is of interest to insurers and insurance intermediaries.

The Guidelines (WP29 [Working Party article 29] guidelines on the Data Protection Officer requirement in the GDPR) clarify that the term 'core activities' refers to the key operations necessary to achieve the main objectives of the relevant business.

The processing of personal data in the context of internal IT services or payroll processing (which are ancillary activities, rather than inextricably linked to the main objectives of the relevant business) does not trigger the obligation to appoint a DPO, according to the Guidelines.

The term 'large scale' is not defined, but the Guidelines note that there are some cases that are surely large scale (e.g., processing at a regional, national, or international level), and some cases that are not large scale (e.g., processing of personal data of an individual patient by a doctor). But, most business activities will fall somewhere between these two extremes. The Guidelines recommend that businesses should consider the following factors in determining whether a given processing activity is 'large scale' or not:

- The number of individuals affected (either in abstract or as a percentage of the relevant population);
- The volume of data, and/or the number of categories of data, being processed;
- The duration or permanence of the processing activities; and
- The geographic scope of the processing activities.

The concept of 'regular and systematic' includes, among other things, tracking and profiling on the internet (e.g., the use of cookies for behavioural marketing purposes). The Guidelines make clear that 'regular and systematic' means any activity that is (i) repeated (with any degree of frequency); and (ii) planned or strategic (i.e., more than an accident or a coincidence).

It is clear that insurance intermediaries, as well as insurance companies, can be included among the subjects who process personal data according to regular and systematic monitoring. The WP29 also intervened to point out that it may be useful to proceed with the designation of the DPO, even where not mandatory, as this helps to improve the 'privacy image' of the owner and/or manager for accountability purposes.

The GDPR represents a burden for insurers, in terms of new tools to be compliant with the new rules but also a challenge, considering the innovative forms of insurance coverage.

References

Cappiello A (2018) Technology and the insurance industry re-configuring the competitive landscape. Springer, New York

Clarke M (2009) The law of insurance contract. Informa Law, London

Forgó N, Hänold S, Schütze B (2017) The principle of purpose limitation and big data. In: Corales M, Fenwick M, Forgó N (eds) New technology, big data and the law. Perspectives in law, business and innovation series. Springer, New York, pp 17–42

Glancy DJ (2012) Privacy in autonomous vehicles. St Clara Law Rev 52:1171–1239

Helberger N (2016) Profiling and targeting consumers in the internet of things. In: Schultze R, Staudenmayer D (eds) Digital revolution: challenges for contract law in practice. Hart Publishing, Baden-Baden, pp 135–161

Hoeren T, Kolany-Raiser B (eds) (2018) Big data in context. Springer, New York

Mahmood S, Sayers S (2015) Connected cars: an approach to dealing with the privacy risks. Priv Data Prot J 15(8):3–5

Marano P, Noussia K (eds) (2020) InsurTech: a legal and regulatory view. Springer, New York

Maurer M, Gerdes JC, Lenz B, Winner H (eds) (2017) Autonomous driving: technical, legal and social aspects. Springer, New York

McGurk B (2019) Data profiling and insurance law. Bloomsbury Publishing PLC, London

Nazzaro AC, Landini S (2020) Blockchain e assicurazioni. In: Valentino D (ed) Commentario del codice civile. Dei singoli contratti. Leggi collegate. Utet, Torino, pp 361–424

Pype P, Daalderop G, Schulz-Kamm E, Walters E, von Grafenstein M (2017) Privacy and security in autonomous vehicles. In: Watzenig D, Horn M (eds) Automated driving. Springer, New York, pp 17–27

Schultze R, Staudenmayer D (eds) (2016) Digital revolution: challenges for contract law in practice. Hart Publishing, Baden-Baden
Schwartz PM (1999) Privacy and democracy in cyberspace. Vanderbilt Law Rev 52:1609–1702

IoT and Privacy

Dianora Poletti

1 IoT in the Current Digital Society

The expression Internet of Things (IoT) is mostly used to describe an ecosystem composed of objects connected to the network through sensors (but also satellites, GPS, microphones, video surveillance, remote-control equipment, etc.) that interface with the physical world and interact with each other, exchanging information on their status and the surrounding environment without the need for human intervention. The neologism (Ashton 2009; Weber 2009) summarises the transformation of the Internet into a responsive connected structure, composed by a crowd of connected objects. These so-called 'smart objects' are capable of gathering, processing, storing and transferring data, from which further information can be processed. For some time, IoT has extended beyond traditional devices such as tablets, computers, or smartphones to potentially include all objects: from everyday ones to implantable ones.

IoT represents a new stage of progression for the Internet. It is a rapidly growing technology, which has already experienced a big increase due to the COVID-19 pandemic and will further explode with the transition to the 5G connection standard. Thus, a new stream of data drawn from smart devices will be generated. The infrastructure has undergone a significant evolution: from the Internet of People to the Internet of Things and then to the Internet of Everything (IoE). In the IoE, the connection is not only between objects (through the 'machine to machine' technology) but also between people and even animals, living together in a connected world in which they 'talk' through any data stored in a database. More precisely, the connections concern objects, objects and individuals' devices, individuals and other objects, and also objects and back-end systems. The physical and virtual world are no

D. Poletti (✉)
University of Pisa, Pisa, Italy
e-mail: dianora.poletti@unipi.it

R. Senigaglia et al. (eds.), *Privacy and Data Protection in Software Services*,
Services and Business Process Reengineering,
https://doi.org/10.1007/978-981-16-3049-1_15

longer distinguishable: the IoT, as recognised by the European Union in the Strategy for the digital market adopted in May 2015, "merges physical and virtual worlds, creating smart environments", and is closely linked to the notions of "pervasive" and "ubiquitous" computing.

The phenomenon has potentially no boundaries, as highlighted by the definition of IoT given in the Recommendation of the International Telecommunication Union ITU-TY 2060 of 2012 (Overview of the Internet of Things), in which the IoT is defined as "a global infrastructure for the information society, enabling advanced services by interconnecting (physical and virtual) things based on existing and evolving interoperable information and communication technologies".

Unsurprisingly, IoT represents one of the key technologies of the current and future information and communication technology sectors. The IoT infrastructure is already functioning and will be implemented in several sectors, both private and public, domestic and industrial: from smart home to smart city, from public transport to agriculture, from Industry 4.0 to healthcare, and from the automotive to commercial network. Gartner estimates there will be 25 billion connected devices on the planet in 2021.

The IoT is therefore at the heart of the transition to a fully digitised society: it enhances the data set to be analysed with artificial intelligence (AI) strategies; it contributes to the creation of Big Data; it uses cloud computing technology and implements the presence of data platforms that interact with other programs and users. At the core of IoT, there are simple tracking devices (like body activity measuring devices) that transmit information to users, operators and more complex networks and environments, such as those that might govern a smart factory or a smart city.

Europe aims to enhance this technology, combined with AI and robotics, considering it decisive to develop the digital market and to strengthen business competitiveness. European Union has long been working to reinforce confidence in the use of IoT and to endorse Europe itself as a world leader in the field, along the way like a path which led to the adoption of the General Data Protection Regulation (Reg. 2016/679/EU, GDPR). The Expert Group Report published in May 2019 was on the basis of the Commission's Report to Parliament and Council on the security and liability implications of artificial intelligence, the Internet of Things, and robotics (February 19, 2020), accompanying the AI White Paper. Furthermore, in October 2020 the European Parliament published a resolution with recommendations to the Commission on a civil liability regime for AI. AI, IoT, and robotics have many features in common—as the report states—because first of all, they allow a combination of connectivity, autonomy, and data dependence. In addition, they generate similar problems on the side of data protection (which will be most considered here), cybersecurity, and consequently liability.

2 Data Flow in IoT

The processing of personal data in the IoT shows peculiar features and functions under different perspectives. First of all, the interconnected ecosphere of IoT appears far from an univocal relationship between the data subject and the data controller. The dynamism of data, due to its uninterrupted flow, results in a bidirectional or multi-directional correlation, either involving the subject and the object and third parties that process the data to improve the functionality of the object itself or involving only the objects (Verseman and Friess 2015). In fact, objects collect information from the subject in order to present other information requested to the latter, even transforming the language of machines into a computer voice, as happens with smart assistants.

In the IoT system, personal data collected in large quantities yield 'fruits', so to say, because they can generate other data (inferred data and metadata). Through the flow of data, the individual is analysed in a given environmental context (at home, at the workplace, on the street) and in their relationship with other individuals, being localised and immersed in their own and others' data. The combination of dat and the use of AI and machine learning techniques allows in-depth profiling of users from several points of view, from consumption patterns to health status, ultimately life patterns. Moreover, machines have acquired the ability to predict behaviours, so as to anticipate (or even steer) certain future choices of the user or groups of users.

In IoT architecture, complexity characterises both subjective and objective features. On the one hand, there are many members of the supply chain that interface with the data: for instance, device and integrated device manufacturers, application developers, software development houses, social platforms, further data recipients, data platforms, and standardisation bodies. On the other, the functioning of 'smart' objects rests on a plurality of components, parts, software, equipment, or services, which constitute the new technological ecosystems. Moreover, there is a need for product updates and improvements after placing them on the market.

As for the type of data collected, personal data, but also special categories of personal data, such as health data (for example, detected by wearable or implanted medical devices or even simple electronic bracelets) and biometric data (retinal structure, iris shape, facial recognition) or even non-personal data, can be taken into consideration (Giovanella 2019). Through this information and its intertwining connections, the IoT environment allows for a particularly invasive monitoring of people's private lives and potential conditioning of their freedom. The huge amount of data collected, in a variety of formats, puts people at the mercy of third parties, through monitoring of objects: Big Data, IoT, and AI make it more difficult to make free decisions, due to behavioural influence and predictive powers. While people are becoming more and more transparent, especially for the Internet's Big Players, the activities of the latter are still very opaque.

It is no coincidence that, already in 2014 under Directive 95/46/EC, the WP29 (Opinion 8/2014 on recent developments in the field of the Internet of Things) indicated that the IoT "poses a number of significant privacy and data protection challenges, some new, some more traditional, but then amplified with regard to the exponential increase in data processing involved by its evolution". Precisely because of the reported risks, in 2016, IoT technology was the subject of the 'Privacy Sweep', a wide-ranging, international investigation—initiated by the Data Protection Authorities belonging to the Global Privacy Enforcement Network (GPEN)— aimed at verifying compliance with the provisions on the processing of personal data in the specific context. Already at that time, the results of the analysis, which concerned a reduced sample of connected objects, highlighted worrying deficiencies in the protection of data subjects. Just to name a few significant outcomes that emerged, it was assessed that 59% of the IoT devices verified did not offer adequate information on how personal data was collected, used, and disclosed to third parties; 68% of them did not release appropriate information on how the data was stored; and 72% did not explain to users how to delete data from the device.

It should also be noted that the IoT systems are partially subject to Directive 58/2002/EC (so-called e-Privacy Directive) relating to the protection of personal data and privacy in the electronic communications sector, which aims to ensure the safeguard of fundamental rights and adheres to the principles recognised especially by the Charter of Fundamental Rights of the European Union. In particular, the Directive implements the fundamental right to privacy with respect to communications, with the result that, while the GDPR refers mainly to Article 8 EU Charter of Fundamental Rights, aimed at protecting the right to protection of personal data, the Directive has as its main objective the respect of Article 7 EU Charter (addressing, by proxy, privacy understood as confidentiality) (Tosi 2019). Therefore, the IoT stands at the crossroads between these two coordinates, revealing that through the network of connected sensors, injuries to both rights can be generated.

3 The Application Limits of the GDPR

The above-described framework highlights the questionable compatibility of the GDPR rules with the IoT systems (Giannone Codiglione 2016).

From a data protection perspective, it can be said that the IoT is now positioned in the middle between the GDPR, which does not properly fit this infrastructure, and Directive 2002/58/EC, which is under review and should be transformed into a regulation, through a process that is showing long lead times due to the relevance of the interests at stake. Indeed, the directive applies to situations where an IoT stakeholder stores information or accesses information already stored in an IoT device to the extent that these IoT devices qualify as 'terminal equipment'.

In order to test the compatibility with some fundamental provisions of the GDPR, it should be noted that the flow and processing of data from the IoT and the multiplicity of actors crowding the overall processing first and foremost weaken the control of the data subjects on their data, and they also compromise the compliance with the principle of purpose (Pizzetti 2018). The increase in the amount of data generated by the IoT and Big Data Analytics and also cross-matching techniques trigger processing for purposes, which are rarely compatible with the ones for which they were collected; purposes that the data subject, moreover, cannot reasonably expect (Mantelero 2012). This results in processing that does not comply with Article 6(4) GDPR unless a new consent is collected or the application of Member State law can be invoked. As exemplified by WP29 in the Opinion 8/2014, examining three areas of the IoT (wearable computing, quantified self, and home automation), a seemingly insignificant data piece collected through a device (e.g., an accelerometer) can be used to infer other information with an entirely different meaning (e.g., individual's driving habits).

Consequently, when processing is based on consent, this is unlikely to be adequately informed and granular, so as to cover not only the category of data collected but also the time and frequency with which it is collected, as required by the GDPR. This is due to the data subjects' unawareness about the multiple uses of their data, and to the fact that sensors and objects are neither able nor programmed to provide privacy policies, not even in the form of standardised icons. The e-Privacy Directive also requires the consent of the data subject, unless access to or storage of the information contained in a device is "strictly necessary in order to provide an information society service explicitly requested by the subscriber or user". The consent requirement in Article 5(3) GDPR primarily applies to the device manufacturer but also to any stakeholder seeking access to the aggregated raw data stored in that infrastructure. It also applies to any data controller who wants to store additional data in a user's device. It should not be overlooked that the owner of an IoT device and the person whose data may be captured and monitored may be different, further extending the requirement to obtain consent.

Also, the proposal for the e-Privacy Regulation particularly values user consent. It should be noted, however, that consent seems insufficient to manage the communications occurring through the IoT systems; it is enough to say that the denial of consent often generates a lock-in effect and that its revocation is not capable of generating a total opt-out from the system of uninterrupted data collection.

The use of non-personal data (subject to the Regulation (EU) 2018/1807, the so-called 'Free Flow of Data Regulation') or anonymous data (excluded from the application of the GDPR) does not shelter the data subject. In fact, anonymisation is particularly difficult, and the possibilities of re-identification of the data subject are consequently high, thus preventing the processed data from losing the qualification of 'personal', not to mention that apparently non-personal data, correlated with the environmental context, can easily release (or even appear as) personal information. Where personal and non-personal data are inextricably linked within a data set, the FFD Regulation is without prejudice to the application of GDPR. Examples of 'mixed' data cited by the Guidance on the Regulation on a framework for the free flow of non-personal data in the European Union (29 May 2019 COM 250 final) include precisely "data related to the Internet of Things, where some data allow assumptions to be made about identifiable individuals (e.g., presence at a particular address and usage patterns)".

As for retention, the circumstance that the accumulation of data allows applications—through the use of AI that learns from the experience of users through machine learning techniques—to offer increasingly accurate services clashes with the provision of limited duration to the achievement of the purpose.

In addition, IoT generates cross-border data flows (through using cloud computing), which are regulated by Art. 44 ff. GDPR. The transfer of data between the EU and the USA, after the invalidation of the Privacy Shield by the European Court of Justice ruling C-311/2018 (Schrems II) is a still open issue, which has been subject to public consultation since November 10, 2020 by the European Data Protection Board. However, it should be noted that 5G technology relies on the key contribution of edge computing, which, unlike the cloud, processes data where the data is produced, rather than in a centralised data warehouse. This will certainly reduce the transit of data, which can be processed and encrypted locally and translated into a secure communication protocol before being sent to the data centre or storage resources shared in the cloud.

Regarding the identification of the subjects of the data processing, the GDPR approach identifies a single data controller (or two or more joint controllers able to determine the purposes and means of the processing), and possibly one or more data processors appointed by the data controllers. This approach, however, embraces a static vision, which is not really compatible with the 'chain of processing' characterising the evolution of the digital society. Within this chain, the data gathering may trigger different processing with their own autonomous, specific purposes, which are connected, in turn, to activities carried out by other data controllers, thanks to the results of such processing. The plurality of subjects involved in the complex IoT ecosystems certainly makes the identification of the data controller challenging, required for data protection compliance and on whose role is modelled the responsibility as shown in Art. 82 GDPR.

Liability for damages resulting from IoT system malfunction also raises specific issues. Even with the difficulties outlined above, Art. 82 GDPR distributes, at least minimally, liability between the two parties involved in personal data processing. Transposed into the IoT infrastructure that rule allows an allocation of risk between those who provide an essential component of the smart object or who intervened in the process of data transmission that produced the data breach. Nevertheless, the use of algorithms applied to large amounts of data and the lack of transparency of the decision-making process makes it more difficult to predict the behaviour of smart products and identify the possible causes of damage. Moreover, due to the fact that in the IoT systems the connection concerns objects, another type of liability may emerge, namely, 'product liability' (Mezzanotte 2019): in case of damage resulting from defective products, Directive 85/374/EEC may apply. Specific attention is dedicated to this Directive by the cited Report on the safety and liability implications of AI, the IoT and robotics, which proposes the modernisation of the Directive precisely to adapt it to the technologies considered. In fact, the Directive needs to adapt the definitions of 'product' and 'defect' to the particular features of digital products and digital elements.

Directive 85/374/EEC is also at the heart of the proposal for a regulation attached to the European Parliament Resolution of 20 October 2020 with recommendations to the Commission on a liability regime for AI (Twigg-Flesner 2021). On a theoretical level, it is possible to distinguish between damages (data losses, profiling, infection by malware, unauthorised access to personal data, intrusive use of wearable devices, or unlawful surveillance) subject to the liability provided for by Article 82 GDPR, and damages caused by a defect in the product or one of its components, which falls under Directive 85/374. On an operational level, the inextricable network between products or services on which this technological environment is based and the processing of data may lead to situations in which a failure of the product (e.g., a home automation device or a wearable) due to an undesirable reaction caused by a connectivity surge also generates at the same time unlawful processing of personal data (e.g., an undue communication to unauthorised third parties). The issue of technology-dependent liability and the adoption of a clear common framework on the matter is now at the heart of Europe's regulatory initiatives (Lohsse and Schulze 2019), as highlighted by the above-mentioned Resolution, which contains a proposal for a regulation unfortunately neglecting the problem of the interference between the two liabilities illustrated above.

4 Privacy by Design and Security by Design in IoT

If it is true that the GDPR, after a long drafting process, was born as already outdated, especially as regards its application to IoT and AI, it cannot be denied that it is very neat legislation, used also as a template by non-European countries for their own legislation in this area. Some of the rules the GDPR laid down are adaptable to the IoT framework, particularly if interpreted with a certain level of flexibility so as to ensure the effectiveness and enforceability of the rights of the data subject and ensure at the same time the evolution of digital technologies.

For example, the chain of processing, neglected about the notion of the data controller, emerges on the rights of data subjects, given that controllers must fulfil certain obligations towards other controllers, with regard to data portability (Art. 20(2)), erasure (Art. 17(2)), and notification of requests for rectification, erasure, restriction of processing (Art. 19).

Moreover, one of the novelties of the GDPR regards Art. 22, referring to the prohibition of treatments based 'solely' on automated decisions and with decision-making effect, as well as profiling activities, as defined by Art. 4(4). This provision can apply to the processing of IoT and AI; although, the effectiveness of this right still clashes with the low quality of the consent of the data subject.

The obligation to carry out a data protection impact assessment (DPIA), provided by Art. 35 GDPR in case of "high risk to the rights and freedoms of natural persons", must be compliant within the IoT system. In this case, there is not even a need for extensive interpretations, given that the rule refers precisely to the specific technological context, as confirmed by the provision of Art. 35(3)(a), which mentions the activities involving "a systematic and extensive evaluation of personal aspects relating to natural persons that are based on automated processing, including profiling, and on which decisions are based that produce legal effects concerning the natural person or similarly significantly affect the natural person".

The principles of privacy by design and privacy by default—which are among the tools selected by the GDPR to prevent data breaches—provide for accountability standards of the data controller and can be invoked by national and European DPA as the most appropriate tools to finetune the protection of personal data with the evolution of information technology. If service providers and manufacturers of objects and machines orient their business strategies towards the respect of personal data from the design stage and consolidate mechanisms of minimisation of processing and anonymous de-structuring of data, then it will be possible to allow subsequent activities and processing that are crucial for the development of AI, which remain outside the scope of the Regulation. The role of certification, which is expressly foreseen by the GDPR, even if not yet fully developed in the member states, can certainly be relevant in order to implement systems that are less risky for the user.

The IoT very prominently raises the issue of cybersecurity, which inevitably intersects with liability. Interoperability and flexibility to the further evolution of smart objects after they are placed in the market, as well as the use of free or open-source software for data processing, can pave the way for dangers to a data subject in the case of products based on AI and IoT, such as cyber-attacks or threats that risk compromising parts or systems of the same infrastructure.

Europe has made many efforts along this direction, starting with the adoption of the now long-standing Directive 2001/95/EC on product safety and ending with the harmonised product legislation that follows the horizontal standards of the 'new regulatory framework' (Regulation (EC) No 765/2008 and Decision 768/2008/EC). The basic principle is safety by design, in which manufacturers should verify in advance, on the basis of product use, the accuracy of data and their relevance to safety conditions. Moreover, as indicated by the Report on the safety and liability implications of Artificial Intelligence, the Internet of Things and robotics, due to the complexity of the product, in the risk assessment phase the manufacturer should also consider reasonably foreseeable misuse by the user, according to a rule extended from the Machinery Directive (Directive 2006/42/EC) that is considered suitable to be applied to a highly technological context. The latest act is represented by the Guidelines for Securing the Internet of Things issued by ENISA (European Union Agency for Cybersecurity) of 9 November 2020, which emphasises the principle of 'security by design'. The Guidelines, assuming the supply chain as a model for the IoT, propose good practices for solving IT security issues: in particular, some of the problems that have been identified concerning the circulation of information and compliance with contractual terms, given that recognition "includes errors in design due to lack of visibility into the components provided by suppliers, or overproduction of a product outside of the boundaries of an established contract".

5 Conclusions

Faced with the IoT infrastructure, as well as other disruptive technologies, academics have criticised the level of protection provided by the GDPR's regulatory framework, as well as by the 'traditional' concept of personal data. The latter is challenged especially by the problem of the ownership of inferred data: on the one hand, a proprietary paradigm has been invoked for the greater protection of the data subject and, on the other, the theory of the commons has been invoked.

This contribution confirms the need to go beyond the protection of personal data when dealing with technologies such as the IoT. In order to effectively safeguard privacy, understood as the protection of citizens' fundamental rights and freedoms, it is necessary to examine the impact that the processing of personal and non-personal data can have in the current digital society. The aim is effective to protect the right to self-determination of individuals, which is the foundation of human dignity.

Relevant in this respect is the Declaration by the Committee of Ministers of the Council of Europe on the manipulative capabilities of algorithmic processes adopted on 13 February 2019: at point 9(a) such document encourages member States to assume their responsibility to address this threat by "considering the need for additional protective frameworks related to data that go beyond current notions of personal data protection and privacy and address the significant impacts of the targeted use of data on societies and on the exercise of human rights more broadly". The Council of Europe does not have the power to adopt binding rules, but its indications should be carefully considered attentively also by EU bodies for the interpretation and application of the GDPR.

Regarding the enforcement of rights, individual protection for the data subject had to be accompanied by collective protection also because the damage, when it occurs, will often be multiplied for entire groups of subjects. The significant imbalance between the parties causes the risk of 'loneliness' and lack of awareness of the individual, who is consequently unable to effectively protect themself. The values of dignity, personal development, and equality vs. the possible manipulation and discrimination that the IoT and the algorithmic logic of machine learning and deep learning are capable of achieving, are values that concern the overall community and, accordingly, require collective control.

Consequently, it will first of all be necessary to enhance and complete the provisions of Art. 80(2) GDPR, which already move in this direction, giving Member States the possibility to provide that any organisation or association, independently from the data subject's mandate, the right to lodge a complaint before the competent supervisory authority and to exercise the rights referred to in Articles 78 and 79, when the processing infringes the data subject's rights.

The creation of a sustainable European economy based on digital development, highlighted by the most recent acts directly involving the IoT, necessarily rests on the release of information and the sharing of data. For this reason, even the debate on the alternative between individual and collective ownership of data perhaps does not seem entirely appropriate.

Data sharing can bring benefits and help build a future for the improvement of our lives. Already in 2009, the European Commission explained that the scope of IoT applications was expected to greatly contribute to addressing today's societal challenges (Mukhopadyay 2014). Indeed, the IoT can generate positive effects in several fields of crucial importance for individuals and governmental actors: from the personal sphere (e.g., assistance to vulnerable people and facilitation of daily activities) to the working and urban environments (efficiency in industry, optimisation of agricultural activities, traffic management) up to global challenges (e.g., health, climate change, security). IoT connectivity features can enable businesses and market surveillance authorities to track dangerous products and identify risks across supply chains, according to the Report on the safety and liability implications of Artificial Intelligence, the Internet of Things and robotics. The same Report also envisages large amounts of data from different sources in the IoT architecture to allow products to self-adapt and thus become safer. For all these reasons, Europe is pushing for sharing data in the digital economy: pooling data means enabling scalability and development even for smaller companies.

From the opposite perspective, though, the use of Big Data will expose individuals to increasing risks, which may grow as the scope expands. Sharing may subjugate individuals to the power and supremacy of technology, leaving them subject to behavioural classifications and predictions, which will increase the concentration of digital information power in the hands of a few digital intermediaries or operators.

If the further development of this technology will safeguard the fundamental values of the person, as Europe invokes, even in the consideration of the unavoidable need for digital development, the balance will strike towards the first perspective. This is clearly emerging in the previous formulation of Art. 2, par. 1 Italian Privacy Code, providing that the processing of personal data should be carried out "in respect of fundamental rights and freedoms, as well as the dignity of the person concerned, with particular reference to confidentiality, personal identity, and the right to protection of personal data". If the IoT and the use of AI were to be inspired by a competition without rules for the maximisation of profit, the second, worrying perspective will prevail.

References

Ashton K (2009) That "Internet of Things" thing. In the real world, things matter more than ideas. RFID J 22(7):97–114

Giannone Codiglione G (2016) Internet of things e nuovo regolamento privacy. In: Sica S, D'Antonio V, Riccio GM (eds) La nuova disciplina europea della privacy. Cedam, Milano, p 131

Giovanella F (2019) Le persone e le cose: la tutela dei dati personali nell'ambito dell'Internet of things. In: Cuffaro V, D'Orazio R, Ricciuto V (eds) I dati personali nel diritto europeo. Giappichelli, Torino, p 1213

Lohsse S, Schulze R, Staudenmajer D (2019) Liability for artificial intelligence and the Internet of Things, Beck - Nomos, München, Baden-Baden

Mantelero A (2012) Big Data: i rischi della concentrazione del potere informativo digitale e gli strumenti di controllo. Il Diritto Dell'informazione e Dell'informatica. 1:135–144

Mezzanotte F (2019) Risk allocation and liability regimes in the IoT. In: De Franceschi A, Schulze R (eds) Digital revolution—new challenges for law, München, Baden-Baden, Beck - Nomos, p 169

Mukhopadyay SC (2014) Internet of things: challenges and opportunities. Springer, Switzerland

Pizzetti F (2018) GDPR e intelligenza artificiale. In: Mantelero A, Poletti D (eds) Regolare la tecnologia: il Reg. UE 2016/679 e la protezione dei dati personali. Un dialogo tra Italia e Spagna. Pisa University Press, Pisa, p 69

Tosi E (2019) Privacy digitale, persona e mercato: tutela della riservatezza e protezione dei dati personali alla luce del GDPR e del nuovo Codice Privacy. In: Tosi E (ed) Privacy digitale, Giuffrè, Milano, p 36

Twigg-Flesner G (2021) Guideline principle for updating the product liability directive for the digital age, ELI. https://europeanlawinstitute.eu/fileadmin/user_upload/p_eli/Publications/ELI_Guiding_Principles_for_Updating_the_PLD_for_the_Digital_Age.pdf

Verseman O, Friess P (2015) Building the hyperconnected society, river publishers series in communications 43

Weber RH (2009) Internet of Things. Need for a new legal environment?, Comput Law Secur Rev 25(6):522–527

Blockchain and Privacy

Giuliano Zanchi

1 Blockchain and Data Protection: A Twofold Relation

Blockchain is one of the most interesting achievements in securing and processing information in the digital world. Since its first introduction in 2008 as a response to the financial crisis, aimed to develop an autonomous and sustainable digital financial system, the blockchain showed its efficiency in processing complex operations according to a pre-determined algorithm and without any human intervention. Beyond its original function to structuring a public transaction ledger of bitcoin, the blockchain rapidly showed its large-scale potential in different areas of economic and social relevance. In particular, the absence of a third-party authority or a State-based regulation to have this system work gathered enthusiasm in the digital arena. Its architecture made of automated and irreversible sequences of actions from block to block has proved to be extremely useful, and it is today used in many sectors of legal relevance (for instance, smart contracts, insurances, construction procurements, and public registers).

The specific recorded chain of 'transactions' and the possibility to have the relevant information verified and confirmed by the consent of the many users of the blockchain determine a new generation of audited data with a high level of reliability. The technological design of the blocks of the chain prevents retroactive alteration of data once they passed to the subsequent block. This seizure of data can be viewed as an interesting means for preserving data integrity during the phases of the processing; moreover, the mass-distributed control of the data can increase the quality and accuracy of the shared information, with clear benefits to the authenticity of the whole process and the precision of the ledger.

G. Zanchi (✉)
Ca' Foscari University, Venice, Italy
e-mail: giuliano.zanchi@unive.it

R. Senigaglia et al. (eds.), *Privacy and Data Protection in Software Services*,
Services and Business Process Reengineering,
https://doi.org/10.1007/978-981-16-3049-1_16

At the same time, the absence of any control by any authorities, neither private nor public, and the irrevocability of the procedure, once it has started, have raised severe concerns on data security and data treatment. The decentralization of the ledger and its almost always public (i.e., not restricted) dimension are conditioned to an unleashed exposure of data, in stark contrast with the orientation of the data protection regulations at the national and international level (De Filippi & Hassan 2018; Ganne 2018; Hacker et al. 2018; Holden &Malani 2018; Laurence 2019; Wright-De & Filippi 2015).

"Technology protects, technology harms" is often said. The relation between blockchain and privacy shows it clearly. The need for a higher and higher level of protection to personal data is strongly perceived today, and the structure of the blockchain might look like a technological defiance of all the achievements in terms of privacy regulations and data protection in the last decades. This is not indeed a new story. Once again, with the blockchain, the privacy of humans is threatened by the digital world, which nonetheless has become an essential part of our personal, professional, cultural, scientific and economic life (Levis 2018; Nelson 2018; Smolenski 2018; Zambrano 2017). We cannot (and we are not willing to) limit the potentials of the electronic facilities, but in doing so we accept to sacrifice part of our private sphere, even beyond what we would like because we do not want to refrain from receiving all the benefits of the technology. The challenge for the law is at first to find a way to better balance the relation between technology and humanity by defending undeniable values without suffocating the progression of the tech-world. But the law can do (and is trying to do) more than just defending humans against the threats and perils coming with the digital technology; the law can use this technology to enhance the traditional legal instruments and create the conditions to make profitable use of the technology to its ends.

The blockchains are a tremendous example of a technology whose correlation with data protection laws is not necessarily of conflict and incompatibility. The potentials of blockchain are compatible with a fair and protective data treatment, though the situation is today still largely out of control.

Let us try to see where we are now then, and where we could get to in a near future.

2 Blockchain and Data Exposure

A very common definition of blockchain explains it as a decentralized, digital, distributed ledger made of single records called 'blocks' through which more transactions made of single decisions recorded in the blocks can be performed. The digital chain of blocks has two essential characters: it does not require any human intervention to work (the whole process is driven by an algorithm) and the sequence of recordings of the blocks cannot be retroactively altered (a timestamp is placed on any transactions in any blocks and the subsequent block contains a cryptographic hash of the previous one). Therefore, the ledger can document the sequence of interconnected transactions and makes the entire process verifiable. Blockchain is not governed by authorities; authentication of the database created by blockchains derives from the cooperation of those involved in the blockchain, who collectively perform a sort of independent audit of the transactions operated via blockchains.

It is clear by the proper structure of the blockchain that data sharing is a crucial feature of this technology. The mass collaboration of users implies a remarkable exposure of data; moreover, the database created at the end of the process is itself the result of a data treatment. The often-public dimension of the blockchain raises the bar of attention to a larger scale and the absence of any managing or regulatory authority of the blockchain leaves the problem without points of reference.

On another perspective, though, it has been stressed that data security is highly protected in such a system where timestamps and cryptography minimize uncertainties and concerns on data leaks or double-spending. In other words, blockchain allows data transfers without the typical problems of reproducibility of digital data in digital environments. The absence of a centralized government of the data flux, which is the usual way to prevent such critical situations, is not here an issue, since the blockchain technology removes the technical conditions for any undesired and uncontrolled dispersion of information throughout the chain.

From a legal perspective, the ledger can record all the exchanges of data with an extremely high level of reliability and can map their origin and when these data has been shared or created, and in doing so it shows to us all its potential as a digital alternative to traditional models of verification and 'traceability' in the development of complex legal acts. We can think, for example, about the conclusion of a contract in an offer-acceptance structure or the succession of formalities in the transfer of property interests. The point is that the same security-by-design of blockchains, which we observe in terms of reliability of data towards the making of a legally relevant process, can be emphasized also from a data protection perspective. By efficiently and permanently recording transactions between two or more parties, this technology can show outstanding ability to secure data, although its common open structure asks for a deeper analysis of privacy-related concerns.

3 The Problematic Regulation of Blockchains and the Example of *Smart Contracts*

Blockchains, like many other digital technologies, are not directly regulated by legislators. There are examples of legal systems dedicating attention to some aspects of blockchains, but a complete and self-sufficient discipline of the legal issues implied in the use of blockchains is missing at the moment in the international scenario. We can easily argue that such a regulation, in the traditional perspective of a discipline made of rules and principles able to offer solutions to the potential conflicts emerging from the use of blockchains, might seem inadequate. More precisely, such a kind of regulation might turn out to be unsuitable for a technology whose rapid development makes it very difficult for any legislators to define a stable set of norms. Moreover, the inconsistency of blockchains to rules created and applied by third-party authorities, together with the genetic a-territorial dimension of the digital technologies, explains why it is unlikely that any legal system engages itself in governing blockchains as such. We must accept that the digital transformation to a certain extent falls outside the scope of State law and international regulations, which can certainly deal with specific issues connected to the use of the digital world, rather than governing its whole functioning.

Consequently, many aspects of legal relevance in using blockchains seek for adaptation of traditional (let us say from now on, *off-line*) rules with all the understandable criticalities originating from this kind of effort. One of the hardest aspects of this transposition of rules and principles from the traditional discipline to blockchain technology lies in the regulation of situations which are unique and incomparable to the offline world.

An example comes from *smart contracts*. They can be defined as agreements that are translated into an informatic language and are executed through algorithms; these algorithms apply the agreed terms and conditions and give execution to the promises based on pre-determined solutions in the event of specific contingencies during the execution of the contract. All these possible situations are measured in a cryptographic code and can be managed automatically; the code provides that in case of a certain condition a certain and pre-determined action or effect shall be produced, with no need for any external intervention, and consequences are automatically activated (the transfer of a sum of money, for example). Once the condition happens and the action or the effect is produced, there is no possibility to get back to the previous step in the execution of the contract; once a pre-determined effect is encrypted in the code, there are no possibilities that a different action or effect is produced other than that specifically provided for. These contracts are 'smart' precisely because they cannot be executed in a different way than that previously agreed by the parties. Its automatism ensures that the parties cannot interfere in the contract performance once they have agreed upon the operational conditions of the contract.

Clearly, smart contracts can work perfectly in a digital environment, but they need appropriate technology to satisfy their operational conditions, and blockchains proved to be a perfect digital framework for them. It is at the same time clear, though, that these contracts are remarkably different from those transactions which are traditionally addressed as such in the legal language.

A smart contract is a self-executable contract, in the sense that it does not require the cooperation of the parties and indeed prevents the parties themselves, once the content has been determined, from contributing in any way to its execution. There are actions of effects of the contract that can vary, but always and only if they have been planned in the algorithm. In this regard, it is said that these digital, self-executed and irrevocable contracts cannot remain unfulfilled.

If we consider what kind of transactions lawyers and legal systems are used to call 'contracts', these cryptographic codes that we define 'smart contracts' are no contracts, or at least they are not contracts in the same legal meaning we still attribute to this term. If the parties of the transaction do not contribute in any way to determining the effects of the contract, then there is no fulfilment just because there are no conducts to be set in compliance with the terms and conditions of the agreement. If there is no fulfilment, since everything is already automatically outlined, there is consequently no execution of the contract, at least to the extent that both fulfilment and execution imply legal acts (conducts) and are not a mere fact, such as the irrevocable consequence of a fully automated process.

Fulfilment and execution of the contract require an activity that in smart contracts is not simply *carried out through* a machine (as it is, for example, in vendor machines), but is *carried out by* a machine (Cornelius 2018; Levy 2017; Mik 2017). The will of the parties is not aimed at the construction of a legal act; it is simply oriented towards the occurrence of a fact, consisting of the inexorable sequence of inputs and outputs of the blockchain. The unforeseen events, the negative externalities which, in the traditional framework, determine the need to adjust the contractual performance, are not here brilliantly resolved, as it has been sometimes suggested; they are simply ignored. Even where the smart contract code provides for different consequences upon the occurrence of what we would call a contractual contingency, in the flow diagram of the smart contract it is a predetermined impulse like any other; they are data that follows other data.

The necessary predetermination of any applicable contingencies does not simply dehumanize smart contracts but qualifies them as an entity of a virtual abstract world, where only what is measurable and scheduled can be of relevance and what is not previously considered is like it does not exist. If the contingencies which are not considered in the code, external to the sequence of the blockchain, do not exist, then the reality where the smart contracts are created is not the same in which off-line (traditional) contracts live. The discipline of contracts, on the contrary, is built upon the idea that a contract cannot provide *ex ante* solutions for all the situations that can happen during its execution. It is not simply a matter of government of the externalities, but foremost it comes from the awareness that such a full *ex ante* control of the externalities is not desirable because it means to prevent any degree of elasticity and adaptation of the contract to unpredictable events of the future. By radically denying the possibility for the parties themselves to diverge from the planned accommodation of their interests, it is pretty much difficult to apply rules and principles of the law of contracts to smart contracts and blockchain technology in general (Druck 2018).

The case of smart contracts and the problematic use of the ordinary legal discipline on contracts is a clear example of how difficult it is to use legal standards which have not been expressly designed and conceived for blockchains.

Similar issues come from the application of general disciplines which are not meant to be applied in specific circumstances or to specific acts (like contracts) but to every situation where the goods and values protected by that discipline are involved. Here too we take into consideration rules and principles that might (and frequently have) not been drafted in accordance with all the specific features of the digital technologies but ask nonetheless to be applied. The application of these regulations does not depend on their degree of adaptability to the digital environment, they apply to any situation which entails the interception of those goods and values they are intended to protect. This is the case of the rules and principles on privacy, whose scope of application inevitably involves blockchains since data are certainly shared and managed when this technology is in use.

4 Rules and Principles on Privacy Matters and the Blockchain: The Case of EU Data Protection Laws

We have already mentioned cases and examples of personal data that can be used, shared, modified throughout the blockchain process, and we have stressed that the protection of these data on blockchain can be seen in different ways, as particularly exposed or strongly protected by its technological framework.

From the standpoint of the European Union, the Regulation (EU) 2016/679 (GDPR) contains number of provisions that can be interesting to analyze in connection to blockchains and this way assessing whether the two-folded perspective above mentioned can be somehow verified (Blechschmidt 2018; Finck 2017; Giancaspro 2017; IBM 2018).

It is worth starting from one of the basic principles of the EU data protection law, the obligation to assess the risks of harming the fundamental rights and freedoms of the data subjects when using any device dealing with personal data of third parties, the so-called 'privacy by design' as defined in the first paragraph of Art. 25 of the GDPR. The risk-based approach of this provision requires controllers to adopt suitable measures to prevent risks from actually occurring, and through "an objective assessment (…) by reference to the nature, scope, context and purposes of the processing" (Recital 76 GDPR). To this extent, Art. 25 of the GDPR is clear in asking controllers to "implement appropriate technical and organizational measures, such as pseudonymization, which are designed to implement data-protection principles, such as data minimization, in an effective manner and to integrate the necessary safeguards into the processing in order to meet the requirements of this Regulation and protect the rights of data subjects".

Privacy by design demands a substantive and tailor-made definition of all the necessary technical safeguards to avoid any violation of the rights and freedoms of the individuals whose data are processed. In other words, it is not the law to define *ex ante* all the standards of data protection, it is up to the controller (and the processor) to plan and organize what is necessary according to a self-assessed analysis, even beyond the mere respect of legal prescription.

When applied to the blockchain, Art. 25 of the GDPR implies the organization of specific techniques capable of implementing data protection rules and principles in the many situations where data are processed. Every transaction in blockchain requires entry data, including potentially personal data, to be verified and confirmed by the participants, who obviously get in touch with the content of these data. The structural sharing of data determines the application of Art. 25 GDPR, at least every time a participant is located within the territory of the EU; and in case of participants outside the EU, Chapter "Personal Data, Non-personal Data, Anonymised Data, Pseudonymised Data, De-identified Data" of the GDPR (Arts. 44–50) shall be applied.

One important element to be considered in this area of application of the GDPR is that the participants cannot be identified. The typical way a participant of a blockchain is involved in its process is through a double system of keys: a public and a private key. The public key is a long string of numbers and letters which has the same function as an address to reach the single participant of the blockchain. Every participant has then a private key, which is a passcode to access the digital identity of the participant and participate in the process together with all the other participants, whose real identity remains therefore undisclosed.

This de-identification system is certainly very interesting from a data protection perspective because it is compliant with the prescription of designing appropriate measures able to secure the participants' identity, which is itself personal data.

Another pillar of the EU data protection regulation is the centrality of the data subject, who must be in complete control of the processing of its data. Among the powers and faculties which pertain to the data subject, there is the possibility to ask for rectification of their data (Art. 16 GDPR) or their erasure (the right to be forgotten, Art. 17 GDPR). But data stored on the blockchain is generally considered incorruptible and not modifiable, and this turns to be a serious obstacle to the exercise of the data subject's rights. It is true that when data is uploaded with a cryptographic hash the data subject can make the personal information as such not accessible to the other participants. But in the case of blockchains where data has been registered without any cryptographic hash (for instance, because sharing the data is necessary for the verification and control from the other participants or because the controller has a legal obligation to keep this data fully accessible and with no time limitations) the exposure of data and their perpetuity in the blockchain create a troublesome risk-based approach in designing due and acceptable safeguards according to the Art. 25 of the GDPR and might create a collision with the fundamentals rights of the data subject (Art. 17 and Art. 18 GDPR).

It is evident that verifiability, not modifiability, stableness and absence of time limits are typical features of blockchains that can clash with the data protection legislation, and in particular with the Chapter "Authorities and Private Companies in Regulating Software Technologies" of the GDPR (Arts. 12–23). And this is one of the main reasons why some authors have argued that data protection laws are a formidable example of an area of law that cannot be reconciled with blockchain and vice versa. The challenge for blockchain technology is to find a way to balance technical means to avoid the public exposure of data necessary to have the blockchain perfectly functioning (the so-called *blockchain compliance*).

Let us consider the right to be forgotten (Art. 17 GDPR). The GDPR gives the data subject the right to obtain data erasure in some specific cases, which are nonetheless very wide and mostly referred to a unilateral decision from the interested person, although the effectiveness of the delisting request addressed to companies (Ausloos 2020; Betkier 2019; Byrum 2018; Jones 2016; Lindsay 2014), such as online portals, proved to be not so easy to get, as the European Data Protection Board underlined in its Guidelines 5/2019 on the criteria of the right to be forgotten in the search engines cases under the GDPR. Similar concerns have been detected in the exercise of this right with social networks (Smith, 2014). In a broader perspective, the digital world shows a certain degree of resistance to the recognition of the right to be forgotten which is supposed to be in contrast with the freedom of expression online.

This is not the place to properly discuss the delicate balance between privacy and freedom of expression online. What here matters is stressing the persisting struggle to have the data subjects' rights satisfied in the digital arena, which is data-based and data-consuming. This is true with blockchain as well.

If we register in a blockchain a document, with a timestamp, which is filled with personal data, we witness a clash between privacy regulation and blockchain (Toth, 2018). One solution to this conflict might be to consider that the radicality of the two opposite directions of privacy and blockchains can be sometimes reduced: blockchain is not necessarily immutable, the right to be forgotten cannot be granted in every situation.

On the first side, a blockchain is supposed to be resistant to modifications of data, for the data, once recorded in a block, cannot be altered retroactively without altering the whole sequence of the other blocks and so the entire chain. It does not mean that the data cannot be 'technically' cancelled once the transaction has moved to subsequent blocks. One thing is to say that there is a strict and inevitable sequentiality throughout the blocks of a blockchain and that the data in one block is the necessary condition to move to the next blocks; one other thing is to say that the spent transactions on previous blocks cannot be cancelled once its function (i.e., to move to subsequent blocks) is completed. If the process has moved on and the sequence cannot be reversed backwards there are no reasons in principle to erase the previous blocks if and when this deletion does not damage the integrity of the chain. The system can be stable and does not collapse if no more useful data is cancelled. This erasure surely limits the transparency of the process and diverges with the typical decentralization of blockchains (we need a centralized decision to erase data), but it can be a useful compromise to balance blockchains and data protection laws.

More recently, there are suggestions to have GDPR and blockchain aligned in an alternative way. It has been argued that it is possible to keep data completely out of the chain, with only its cryptographic hash in the blocks. In this case, the integrity of the chain is maintained and there is no data exposure, not even for a limited period. This solution, which has been considered technically and practically feasible, is said to have the advantage of not sharing data in the chain, more than minimizing the risk of violation of rights and freedoms of the data subject. But the advantage is only apparent: as some authors have correctly underlined, it is logically impossible to have all the data erased because we need those data to process the blockchain and the cryptographic hash of personal data is itself a personal data under EU data protection law. The problem is moved from one kind of data to another; the codified version of data can be indeed more easily and efficiently managed, but we could not assume that it implies the absence of any privacy concern.

On the other side, art. 17 GDPR itself provides for some exceptions to the right to be forgotten in the situations listed in its third paragraph. Should this be the case, the blockchain can be maintained as is with all the data untouched. These cases, though, are very specific and not so easy to verify when blockchains are concerned (exercise of the right of freedom of expression and information, compliance with legal obligations or public orders, in case of public interest in the area of public health, archiving, scientific, historical or statistical purposes, establishment, exercise or defence of legal claims).

It must nonetheless be remembered that in many concrete situations, privacy issues are by definition absent and data protection regulations are not relevant, as it happens when blockchain relates to commercial transactions between legal entities not involving any individual's data, as frequently happens in the international trade.

5 Conclusions

We can get back to the statement at the end of the first paragraph and ask ourselves where we are in terms of the evolution of privacy and blockchain. The two terms of this relation are not always settled down in the same way: sometimes the concrete functioning of blockchains requires data exposures that are incompatible with privacy regulations, and the GDPR in particular, while some other times blockchains look suitable—much more than other technologies—to protect individuals' data. It has been said that blockchains and privacy are actually pursuing the same goal, although through different mechanisms: giving individuals more control over their data.

In any case, the legislations are still not equipped to cope with all the challenges of this technology and of course, this technology, like any other, will evolve continuously making it more and more difficult for the legislators to keep up with the new frontiers of the digital world. But it is not possible for the law to ignore it or just fight it. It is necessary on the contrary that the legislators adopt a neutral and proactive attitude towards technologies like blockchain. It is up to the law to take advantage and make the best use of it, reverting a common assumption: technologies are neither a threat nor a benefit, they are what we are prepared to do with them.

References

Ausloos J (2020) The right to erasure in EU data protection law. Oxford University Press, Oxford

Betkier M (2019) Privacy online, law and the effective regulation of online services. Intersentia, Cambridge, pp 183–238

Blechschmidt B (2018) Blockchain in Europe: closing the strategy gap. https://www.cognizant.com/perspectives/how-and-why-europe-sees-blockchain-as-a-way-to-level-the-global-playing-field?utm_source=organic_twitter&utm_medium=social&utm_campaign=Thought%20Leadership&utm_term=na&utm_content=Digital%20Systems%20&%20Technology%20DST.Blockchain.Thought%20Leadership&sf95957450=1

Byrum K (2018) The European right to be forgotten: the first amendment enemy. Lexington Books, London

Cornelius KB (2018) Smart contracts and the freedom of contract doctrine. J Internet Law 22(5):3–11

De Filippi P, Hassan S (2018) Blockchain technology as a regulatory technology: from code is law to law is code. First Monday 21(12). https://doi.org/10.5210/fm.v21i12.7113

Druck JA (2018) Smart contracts are neither smart nor contract. Banking & Financial Services Policy Report 37(10)

Finck M (2017) Blockchains and data protection in the European Union. Max Planck Institute for Innovation & Competition Research Paper No. 18–01. https://papers.ssrn.com/sol3/Delivery.cfm/SSRN_ID3119584_code1137858.pdf?abstractid=3080322&mirid=1

Ganne E (2018) Can blockchain revolutionize international trade? WTO Publications, Geneva

Giancaspro M (2017) Is a 'smart contract' really a smart idea? Insights from a legal perspective. Comput Law Secur Rev 33(6):825–835

Hacker P, Lianos I, Dimitropoulos G, Eich E (2018) Regulating blockchain: techno-social and legal challenges. Oxford University Press, Oxford

Holden R, Malani A (2018) Can blockchain solve the holdup problem in contracts? University of Chicago Coase-Sandor Institute for Law & Economics Research Paper No. 846. https://papers.ssrn.com/sol3/Delivery.cfm/SSRN_ID3093879_code285372.pdf?abstractid=3093879&mirid=1

IBM (2018) Blockchain and GDPR. IBM Corporation, White Paper. IBM Security

Jones M (2016) Ctrl + Z: the right to be forgotten. New York University Press, New York

Laurence T (2019) Introduction to blockchain technology. The many faces of blockchain technology in the 21st century. Van Haren Publishing, Hertogenbosch

Levis J (2018) Countries leading the blockchain innovation movement. Available at tokentarget.com/countries-leading-the-blockchain-innovation-movement

Levy KEC (2017) Not street-smart: blockchain-based smart contracts and the social workings of law. Engag Sci Technol Soc 3:1–15

Lindsay D (2014) The 'right to be forgotten' in European data protection law. In: Witzleb N, Lindsay M, Paterson M, Rodrick S (eds) Emerging challenges in privacy law: comparative perspectives. Cambridge University Press, Cambridge, pp 290–337

Mik E (2017) Smart contracts: terminology, technical limitations and real world complexity. Law Innov Technol 9(2):269–300

Millard C (2018) Blockchain and law: incompatible codes? Comput Law Secur Rev 34:843–846

Nelson A (2018) Cryptocurrency regulation in 2018: where the world stands right now. bitcoinmagazine.com/articles/cryptocurrency-regulation-2018-where-world-stands-right-now

Smith K (2014) The right to be forgotten: legislating of individuals to regain control of their personal information on social networks. Reinvention: an Int J Undergraduate Res 7(1). http://www.warwick.ac.uk/reinventionjournal/archive/volume7issue1/smith

Smolenski N (2018) Blockchain in government. Learning machine. www.learningmachine.com/wp-content/uploads/2017/07/Blockchain-in-Government-2017-Q3.pdf

Steinbeck D (2018) How new EU privacy laws will impact blockchain: expert take. cointelegraph.com/news/how-new-eu-privacy-laws-will-impact-blockchain-expert-take

Toth A (2018) Will GDPR block blockchain? www.weforum.org/agenda/2018/05/will-gdpr-block-blockchain

Wright A, De Filippi P (2015) Decentralized blockchain technology and the rise of lex cryptographia. https://papers.ssrn.com/sol3/Delivery.cfm/SSRN_ID2580664_code2373233.pdf?abstractid=2580664&mirid=1

Zambrano R et al (2017) Blockchain. Unpacking the disruptive potential of blockchain technology for human development. http://hdl.handle.net/10625/56662

Enhancing Transparency of Data Processing and Data Subject's Rights Through Technical Tools: The PIMS and PDS Solution

Alessandro Bernes

1 Increasing User Awareness in the Era of Datafication

The question of whether to provide personal data while interfacing with digital services, such as e-commerce, social media, and e-government platforms, is a false alternative. The lack of access to the digital environment usually degenerates into a social self-exclusion and the loss of benefits, in addition to higher costs. Despite the different lawful bases for processing, everyday activities whirl around a large-scale collection, computation, and sharing of personal data, whose operations are performed by using both software and hardware technologies, such as web services, cookies, apps, and the Internet of Things (IoT). Thus, thanks to Artificial Intelligence, machine learning, and Big Data, the value of personal data has been constantly increasing from a social and economic perspective.

However, individuals are not able to evaluate in practice the consequences of data choices and claim an acceptable right to self-determination. The well-known dilemma of the 'privacy paradox' is enlightened: everyone cares about privacy and data protection but, at the same time, less attention is paid while making information available online, especially when suppliers and traders offer seemingly 'free' online services, i.e., in exchange for data monetisation in other markets due to advertising revenues.

It is remarkable that there is a 'technological gap' that affects the user control of data rather than the lack of regulation. In short, what is currently missing are widespread tools for displaying, for example, which data, originating from multiple sources, have been collected over time, the consents that are granted for obtaining digital products or for marketing purposes, as well as the technical means for monitoring further utilisation of information relating to individuals performed by third

A. Bernes (✉)
Ca' Foscari University of Venice, Venice, Italy
e-mail: alessandro.bernes@unive.it

R. Senigaglia et al. (eds.), *Privacy and Data Protection in Software Services*, Services and Business Process Reengineering,
https://doi.org/10.1007/978-981-16-3049-1_17

parties. The absence of those instruments prevents data subjects from fully understanding both benefits and threats of the ongoing data processing, in so far that, the structural (information) asymmetry existing between digital providers and their users is enhanced, whose degenerations leads to the so-called 'surveillance capitalism' and its outcomes, such as profiling and personalisation without knowing.

The crunch is if the same or different technologies aimed at data processing might be used for rebalancing this asymmetrical relationship, by providing not only better data management but also increasing user awareness over personal data flow. The reference is mostly intended—especially for computer scientists and engineers—to the so-called 'Privacy Enhancing Technologies' (PET), which since the 1990s have been invoked to strengthen a new approach to data protection, known as 'privacy by design'. Accordingly, PET stays for "software and hardware solutions, i.e., systems encompassing technical processes, methods, or knowledge to achieve specific privacy or data protection functionality or to protect against risks to privacy of an individual or a group of natural persons" (ENISA 2015).

The concept of PET is very broad and the existing models are countless. Several techniques of data anonymisation are actually performed by technical means (k-anonymity, differential privacy), as well as secure multi-party computation, data security measures (homomorphic encryption), data minimisation methods (attribute-based authentication credentials), safe online browsing, default privacy preferences settings, etc. Furthermore, some tools or patterns permit the general principles of data processing to be implemented in an operational way, as well as ensuring an effective exercise of the data subject's rights.

With the advent of the digital age, the right to data protection cannot be verged to barely devise a defensive logic against this new technological revolution. On the contrary, the protection of individuals and the tools for data processing have to merge so that law and ICT can be considered both for their regulative side and technical aspects. In this respect, the European Commission considers PET complementary to the General Data Protection Regulation (EU) 2016/679 (GDPR) for defining a stronger data protection framework (European data strategy 2020).

The following analysis will focus on how user-centric technology empowers individuals to be confident about operations that involve data processing. This represents also an important trigger to increase transparency and trust of digital providers, to avoid the opposite perspective of undue profiling or excessive data collection in context-dependent situations, which ends up with resorting to fully ad-blocking tools or, which is worse, by escaping completely from the online environment. Moreover, particular types of PET might also limit monopolistic drifts of the data-driven economy, i.e., the potential abuses resulting from detaining huge amounts of information about individuals without real awareness.

The role played by law within the renewed interest of PET is to guide and evaluate how data protection may be better implemented throughout technologies, especially concerning transparency-enhancing ones. Therefore, the right to obtain information about the processing of personal data shall be considered with computer science and engineering studies, while jurists should provide specific 'guidelines' for the practical implementation of software to all those who have to deal with data processing, not only for demonstrating compliance to GDPR but also enforcing individual rights and freedoms through technologies. Hence, legal principles and rules must ensure that human beings are settled at the centre of attention, by considering, however, what are consumer choices in the operations that are performed on personal data.

2 A Matter of Transparency: the PIMS Technical Standard

The 'hyper-connected life' in which we live has brought a substantial change in daily activities, as the neologism 'onlife' has well sketched the blurred distinction between reality and virtuality (Floridi 2015). Thus, it is hard to draw a line between the informational economy and the social system as a whole, moving from the massive collection of personal data performed by digital providers to the warped usage of information for 'influencing' consumer behaviour, sometimes apart from commercial purposes. With the above in mind, the continuous disclosure of information, as the COVID-19 pandemic has clearly shown, has proportionally decreased the user perception of the data processing carried out through smart devices, especially where consumers have no real alternatives to the service requested and no genuine choice is given.

Nevertheless, individuals lose control of their data even if the lack of transparency is not intentional. The widespread 'notice and consent approach', which requires the approval of long but also 'vague' written documents over data processing, is an inefficient model for guaranteeing the right to self-determination in the digital world, considering that most people do not read the forms on a regular basis and accept them without real awareness. The problem leans on how to explain to data subjects in a clear and intelligible way the processing of personal data. Consequently, individuals are not able to reconstruct their digital identity, today more fragmented than ever on the Web. For example, users are often unable to graphically view (and gather at once) which data are stored by making online purchases, ordering food for home delivery, and so on.

The same happens in the case of cyber-attacks. As laid down by Art. 33 of the GDPR, when a personal data breach puts the rights and freedoms of natural persons at high risk, the data controller must inform, without undue delay, the data subject, describing the categories and the approximate number of the personal data concerned. However, it is difficult for users to understand the seriousness of the leaks because they are not fully aware of the quantity and quality of personal data previously collected.

Yet, even if data subject's rights have been boosted under the GDPR, the requests of access to data, to obtain copies, and to erase certain information (if they are aware of it), which are submitted to the data controllers, still remain slow and cumbersome. Individuals are less convinced to act, since sending just a single email seems, paradoxically, detrimental.

What is missing here is not the development, but rather the rapid adoption of technical tools, which could permit, in a simple and clear manner, to increase user awareness over data flow across the digital world. The static representation relating to data processing should move towards a dynamic inclusion of the data subjects for counter-balancing their role. Otherwise, the compliance with the GDPR by the data controller may be improved, as well as the company's reputation in the digital market.

The solution proposed with the Transparency Enhancing Tools (TET) usually falls into the model of the Privacy Information Management Systems (PIMS), also called Personal Data Management Systems (PDMS), which consist of specific products and software services undertaken by the data controllers, while the data subjects directly face the processing of personal data and better control their information on online activities. Hence, PIMS include, *inter alia*, tools to decide, at a granular level, which data are processed, for what purposes, for how long, and then choose among different levels of data sharing. In particular, it shall be made possible to provide a dynamic display of the real-time data processing (e.g., proposing an intuitive and user-friendly 'dashboard'), involving also data breach notifications, as well as to enable management mechanisms for an easier exercise of the data subject's rights through the implementation of simple 'point and click' solutions. Furthermore, it should be possible to let users deal with risk assessment, by choosing the categories of recipients and the types of data further processed for other purposes, adopting stronger security measures for special categories of data (e.g., strong authentication, client-side encryption), as well as deleting particular contents, no longer relevant for the data controller or the data subjects themselves. Other specific functions, such as the possibility of storing data directly on user devices or on a cloud-based repository and even performing data operations locally may be added to reinforce users' empowerment, as will be seen later.

More broadly, not only PIMS try to rebalance the information asymmetry between the two different figures existing in the digital market, i.e., the professional and the consumer, but also data subjects are provided with the practical means to adequately verify and deal with the processing of personal data. Consequently, individuals are moved from being passive data sources to become 'proactive' parties of the data economy.

Nevertheless, in order to strengthen the protection of individuals within the digital environment, the approach of the data controllers is critical, while choosing adequate technical and organisational measures and accountable safeguards in relation to the specific context. In this respect, much greater transparency must not lead to the opposite of an information 'overload', especially in the context of automated processing and profiling: too granular information (and features) do not deem beneficial for data subjects (and data controllers too). The criteria that the controller should follow to structure the information disclosure shall be compliant with the principle of necessity, proportionality, and reasonableness. In this sense, particular attention should be paid not only to the purposes of data processing but also to the preferences of those who request the digital service. The acceptability and usability of the customers are fundamental elements to be taken into account.

In a nutshell, the principle of transparency should not be seen at the level of whether certain information over data processing is received or not, which remains mandatory to be provided to the data subject, rather on how to better display a well-structured representation of the ongoing data processing. Thus, with having different levels of accuracy about the information, such PIMS should represent a scalable solution more focused on the user. In this sense, data controllers need to adopt appropriate software systems, tools, and technologies concerning the characteristics of the processing carried out, which are strongly linked to data protection by design and by default settings.

3 A Matter of 'Design': How to Choose Transparency Enhancing Tools

Although Art. 25 of the GDPR does not formally embrace the design and the development of new products and applications (Recital 78), data protection by design and by default requires the controller to choose the types of software (and also hardware) solutions that will be used in the following operations. In this light, systems such as PIMS should be investigated for the fulfilment of GDPR obligations from the initial stage of the processing—even before the time of the determination of the means, and whatever technical measures will be addressed to the future relationship with the data subjects—with particular attention to which data are "adequate, relevant, and limited to what is necessary in relation to the purposes for which they are processed ('data minimisation')" (Art. 5(c) GDPR). The functionalities and features available on those measures may affect not only the time of data collection; on the contrary, such tools require a continuous update (e.g., security, data accuracy, retention period, etc.), in order to cover the entire lifecycle of the processing.

More broadly, PET are not only security options in the context of operations carried out with personal data; instead, they represent a precise strategy, from both technical and organisational perspectives, for integrating tools respondent to general principles of data processing. In other words, the structure of the processing should be combined with the function to which those technologies are addressed, to guarantee, on one hand, the fairness and the transparency of data processing and, on the other hand, to enable the data subjects to exercise the rights that have been recognised to them.

The proposed solution relating to a higher level of transparency over data processing is deemed beneficial at overtaking not only the 'technological gap' of user awareness existing on the demand side but also is aimed at enhancing both data protection and trustworthiness of the online environment, as well as allowing digital providers to better dealing with their customers.

Nevertheless, the technological approach aimed at data protection allows also a fruitful dialogue between technology and law, although there is a widespread belief that the GDPR has merely acknowledged a digital neutrality understatement, according to which legal rules do not recommend the adoption of a certain technology. If this is true, it should be considered that appointing a specific tool may lead to the ineffectiveness of the legal framework or potentially overtake the legal discipline by shaping the behaviours of individuals while interfacing with digital tools. What matters most is not which rules or technologies have to be taken; the crucial issue is to identify a functional relationship between the purposes and the means according to the relevant data processing about the selection of a certain technology.

Let us say, for example, that a digital service provider in its mobile app implements a feature in which it is possible to screen a 'flowchart' that shows what personal data have been collected, the current data usage, the subsequent dissemination, and then reuse for purposes other than the initial processing. The variety of technical components that can be added to achieve this functionality is quite different, but some of them remain more affordable than others, at least in terms of cost of implementation and by considering the nature, scope, context, and purposes of the processing, as well as the risks for individual rights and freedoms (Art. 24 GDPR).

(a) An effective representation of either personal data or data processing lies, first and foremost, on the user-friendly interface and specifically on graphics. Thus, the textual form or a simple list of data collected seems a not so efficient solution. Differently, it looks appropriate to group data by different semantic categories, while also using colours, standardised icons, diagrams, percentage elements, and time intervals, along with a brief description of each type of data. This may provide a clear overview of the intended processing and let users make meaningful choices, including behavioural nudges over data disclosure, such as profiling and personalised services.

(b) The adoption of a layered approach is very useful, where the most relevant information to be obtained by the data subject is visualised at a glance and, over time, enables further individual interaction. In particular, user-centric solutions should be made available to those who are even more affected by the 'digital divide', such as minors or elderlies. Additionally, many data subjects may not be satisfied with a minimum level of knowledge so that a 'Chinese's boxes' mechanism, with different pop-ups/windows that can be opened in sequence, should be offered to users who expect to receive more detailed information over data processing.

(c) A notification system certainly allows data subjects to notify users of the processing of personal data performed by public authorities or on the ground of legitimate interest, as well as to comprehend the reuse of their data after the collection. Otherwise, an alert/warning could inform the user of detected threats, potential harms, or critical exposition to possible data leakages for having overloaded the digital service with too many specific data, as well as, in the case of information excess, the same tool may suggest an action for erasing older data, which are not even useful for the profilers too.

(d) Another feature is to provide a sort of intuitive control panel, such as a 'dashboard', that constitutes a highly customisable and attractive interface through which one can communicate in real-time with the data controller, as a kind of 'personal data customer service', embedded in the digital service's smart app. This would also enable a 'technological enforcement' of data subject's rights. Where technically feasible, end users may also grant, through default software preferences, consent to data processing for multiple purposes across different online service providers ('sticky policies' mechanism).

In essence, data protection by design does not require that the legal constraints relating to the processing of personal data shall be translated into algorithmic packages or in Boolean terms but only considered by data controllers when choosing technical and organisational measures to be implemented for a better comprehension of the users over the processing of personal data and, in this way, to ensure an effective data protection regime. The alternatives for technical tools, such as PIMS, are undoubtedly influenced by the rules and principles relating to the processing of personal data.

While interfacing with data protection by design and by default, an assessment of the concrete situations under which the general principles relating to the processing of personal data apply is always required, by taking into account the 'input' and the 'output' that rely on the adoption of a certain technology through a risk-based approach. This is also an essential part of data processing and, consequently, of technological regulation, while making the tools (and algorithms) being, at the same time, testable and contestable (Hildebrandt 2017). In this sense, codes of conduct and certification seals (e.g., ISO 27,701 on PIMS), which involve various combinations of technical tools, adequate safeguards, and privacy patterns, are also essential to define the architecture of PET. Moreover, specific guidelines by the national and European supervisory authorities shall be attempted.

As happened for the debate on *Lex Informatica*, technology may condition the law, by facilitating certain solutions, but rules must not lose their function of regulating the development of new digital tools. Instead, legal discipline might assure to verify whether the effectiveness of the protection of the individuals would be achieved, due to the complexity and fast changes in today's reality, by selecting the appropriate technologies in relation to the ongoing data processing not in the abstract but in the context.

4 PDS: A Winning Model?

Providing the data subjects with practical means for interfacing with the processing of personal data might help to increase their awareness over data flow in the digital era, as well as empowering their autonomy in informational capitalism. In this respect, Personal Data Stores (PDS), known also as personal data clouds or data spaces, could be seen as a cluster of PIMS that are specifically based on user active participation. Specifically, PDS could be seen as an alternative to the current business model of data aggregation into centralised servers and proprietary lock-ins, in which data remains under the direct control of a few digital service providers and traders ('silos'), and further information is used for making consumers take some decisions. In other words, the issuance of PDS would act against the current structure of the Big Data economy, in which only a few monopolists can afford to offer the services 'for free' and later monetising personal data from their 'position rent' of having individual profiles.

Therefore, the solution offered by PDS conveys an identity provisioning service and a distributed space where data of a person can be gathered at one place. Consequently, data processing is enabled according to the selective choices of the individuals, by putting them in the position to decide by whom their data are processed without unnecessarily replicating them, as well as to perform remote computation safely on the single device with user intervention.

The idea of having one or more trusted data custodians is currently reincarnated in the imaginaries of PDS (Lehtiniemi 2017), for which individuals may be considered better as 'holders' (not owners) of their data, meanwhile, the various digital providers can only access personal data selectively, depending on the will accorded by users on a granular basis. Furthermore, any type of data (e.g., contact information, photos, files, etc.) can be stored locally on the user's device or a cloud repository, powered by a third party who acts as a 'infomediary'. Here software (and also hardware) technologies provide data subjects with a decentralised approach to data processing, in terms of intermediation between digital suppliers and consumers and away from the current data-centric modus operandi (see for example Solid web decentralization project).

PDS can be seen as a user-friendly tool to give insight and much control over data that have been collected by various information society services—considering that consent management is usually set at an early stage of the processing and by a one-off decision rather than continuously—as well as for keeping data updated and performing analytics and risk exposure audit tests. Furthermore, the PDS usage may solve the problem of international data transfer beyond EU borders, by ensuring data storage under the GDPR.

Yet, PDS shall act as a medium to easily exercise the rights which have been recognised to data subjects under the GDPR, especially the right of access, the right to erasure, the right to rectification, the right to object, the right not to be subject to a decision based solely on automated processing and the right to withdraw of consent to the processing. Moreover, the presence of a single point of access, compatible APIs, and standard protocols leads to data movement between different systems in a machine-readable format, as it allows to switch from one digital service to another, making the right to data portability effective.

Nevertheless, PDS architecture raises some concerns under the GDPR. Among others, it is not clear the role (and the liability) of those 'infomediaries', which may act as data processors nominated by digital platforms, as well as autonomous data controllers, depending on the case. Since users cannot be treated as data controllers themselves for service usage, a possible solution could be found in the joint controllership, in which PDS suppliers and digital providers arrange their respective responsibilities by contractual agreement. Yet, although consent should not be seen as the only legal basis for information society services—especially for the performance of a contract or in case of a prevailing legitimate interest—another issue is how to grant in practice a valid consent at any data request from different data controllers with sufficient awareness, while also considering different levels of access and data usage.

Invoking software solutions, such as user privacy default preferences and settings (as in the vision of the Semantic Web), periodically updated, perhaps could lead to proper information.

The promising market for PDS is even more challenging. The practical implementation of those technologies needs to overreach today's company-centric model, based on the presence of entry barriers and competitive advantage. However, much greater transparency of data processing should lead to more accurate and up-to-date data repositories so that digital companies could perform their activities with a more accurate dataset, either for individual or group profiling, while users may have more propensity to supply their data in order to achieve more detailed targeting. This would shift the current business model from a siloed-centric approach to the quality of the offered service as a stimulus to secure more customers, by also considering the 'network effect'. Again, the company's reputation certainly would be improved.

Anyways, it is still blurry who has to pay for the ecosystem. The users cannot be asked to pay for enhanced data protection, but, at the same time, it is not worth it for them that they receive a pecuniary compensation for any use of their information, embracing 'data ownership' instead of fair data usage. Otherwise, for activities in which the core business is not represented by the collection of personal data, the adoption of PDS would be seen as a major cost of compliance whether data users are charged for entrusting to a third party, even if data controllers would be less exposed to risks, as they no longer have to manage a huge amount of data directly.

Through PDS, the accountability to which the data controller is asked leads to the adoption of a 'castle of glass' architecture, which is also respectful of the legal requirements for data processing; on the other hand, the information disclosure towards the data subjects is simplified, as for the better quality of the service offered. Therefore, the principle of transparency is the trigger of data protection since it acts not only as a connector for the various operations over personal data, but also, by coordinating the purposes and means of the processing, it enhances the effective exercise of the rights that the GDPR has granted to the data subjects and, in this way, it strengthens the enforcement of the individual rights and freedoms. In the end, PDS could be used to assist in rebalancing the information asymmetry existing between data controllers and data subjects, by enhancing data subjects' empowerment and users' involvement. Hence, having a widespread PDS commercial readiness, at the time of deciding the means of the ongoing processing, can be foreseen as a choice of 'transparency by design'.

5 Transparency as a 'Tool' for the Democratic Regulation of Technology

As shown above, Transparency Enhancing Technologies could be used for enabling users to keep track of the processing of personal data online both from an *ex ante* (obtaining information on data processing) and *ex post* (exercising the data subject's rights) perspectives. In particular, those tools could solve several issues arising from the IoT ecosystem, where it is often difficult to have appropriate knowledge relating to data processing, due to the small size or even the absence of a screen through interfacing with the digital service and displaying which data are incrementally collected. In so far that the presence of a dedicated PDS allows not only to empower the user leading to more active data management but also to show how data flow is inherent to the physical devices.

However, data protection by design is not a rigid concept since it needs a careful assessment not only of the risks for individual rights and freedoms but also for the context in which the processing of personal data takes place. In essence, the effectiveness of personal data protection measures is not relinquished solely to the legal framework or vice versa to technical measures: the necessity and proportionality tests have to take into account the present and future technical solutions, by considering also the interconnection between legal rules and ICT support.

End-user preferences need to be taken into account for enhancing the capacity of TET to be deemed beneficial for stronger data protection, as well as for fostering the trustworthiness of the (future) data-driven economy. In this respect, PIMS/PDS samples could be seen as an opportunity to intervene in 'surveillance capitalism', by giving citizens an active role within the digital world, thus allowing them to face the monopolistic drifts of undue profiling, AI, and machine learning biases, as well as increasing consumer confidence on digital services. The solution must be sought in the democratic choice of the technology to be used, by rejecting the threats of the 'algorithm dictatorship'.

The outbreak of a market of such technologies, which allows better control of personal data, while the individuals are placed at the centre through an inclusive and non-alienating approach, allows them to make more informed decisions in the era of datafication. If the architecture of the PIMS/PDS is built on a personalised and human-friendly basis, i.e., by adopting an easy-accessible user interface (through standardised icons, colours, or diagrams), not only the respect of the GDPR but also the mutual trust between the digital service providers and their customers can benefit. In this respect, the EU Digital Single Market could be seen as an alternative data economy model, where the industrial policy may boost PET based on transparency, designed in Europe, and exported around the world.

As a virtuous circle, the adoption of specific technologies for data processing could enable the data subjects to better trust the services proposed in the digital environment, as they are offered more transparently. Empowering users through technologies would support the assessment and the control of the behaviour of private companies and public bodies, by fostering an active contribution of both consumers and citizens for the private enforcement, as a complement of the public regulation of data protection. This also strengthens the coordination between the rules set out for data protection, consumer protection, and competition in the digital market (see also the Commission Proposal for an EU Digital Governance Act announced in November 2020).

What is still missing for the widespread adoption of PIMS/PDS is perhaps the lack of consideration of the users of the security threats of the Network, as well as their responsibility on the Internet. In this respect, the e-government sector could be seen as a promising sector for the implementation of such tools (especially for public eHealth and eID-ready services), as well as for successful information campaigns that could certainly increase the doubts raised from the usage of those new data processing technologies.

In the digital age, countervailing powers that are to be considered effective require to be supported by the same technologies that permit the processing of personal data for the traders and online service providers. On the other side, new technologies should be presented as accessible and open source, by letting the community control them as an essential element in the information society.

As Recital 4 of the GDPR affirms that "the processing of personal data should be designed to serve mankind", the same goes also for the use of technology, which does not need to be neutral but has to be designed and developed to help humans.

References

Article 29 Working Party (2018) Guidelines on Transparency under Regulation 2016/679. https://ec.europa.eu/newsroom/article29/document.cfm?action=display&doc_id=51025

Bravo F (2019) L'«architettura» del trattamento e la sicurezza dei dati e dei sistemi. In: Cuffaro V, D'Orazio R, Ricciuto V (eds) I dati personali nel diritto europeo. Giappichelli, Torino, pp 775–854

Bygrave LA (2017) Data protection by design and by default: deciphering the EU's legislative requirements. Oslo Law Rev 4(2):105–120

Crabtree A, Lodge T et al (2018) building accountability into the Internet of Things: the IoT Databox model. J Reliable Intell Environ 4:39–55

European Commission (2020) Communication from the Commission to the European Parliament, the Council, the European Economic and Social Committee and the Committee of the Regions. A European Strategy for Data, 19 February 2020, COM/2020/66 final. https://eur-lex.europa.eu/legal-content/EN/TXT/PDF/?uri=CELEX:52020DC0066&from=EN

European Commission (2015) An emerging offer of "personal information management services". Current state of service offers and challenges. https://ec.europa.eu/newsroom/dae/document.cfm?doc_id=40118

European Data protection Board (2020) Guidelines 4/2019 on Article 25 Data Protection by Design and by Default. Version 2.0. Adopted on 20 October 2020. https://edpb.europa.eu/sites/edpb/files/files/file1/edpb_guidelines_201904_dataprotection_by_design_and_by_default_v2.0_en.pdf

European Data Protection Supervisor (2020) Opinion 3/2020 on the European strategy for data. https://edps.europa.eu/sites/default/files/publication/20-06-16_opinion_data_strategy_en.pdf

European Data Protection Supervisor (2016) Opinion 9/2016 on personal information management systems. Towards more user empowerment in managing and processing personal data. https://edps.europa.eu/sites/edp/files/publication/16-10-20_pims_opinion_en.pdf

ENISA (2017) Privacy and security in personal data clouds. https://www.enisa.europa.eu/publications/privacy-and-security-in-personal-data-clouds/at_download/fullReport

ENISA (2015) Privacy by design in big data. An overview of privacy enhancing technologies in the era of big data analytics. https://www.enisa.europa.eu/publications/big-data-protection/at_download/fullReport

Finocchiaro G (2012) Riflessioni su diritto e tecnica. Diritto Dell'informazione e Dell'informatica 4–5:831–840

Floridi L (2015) The Onlife Manifesto. Being Human in a Hyperconnected Era, Springer, Cham

Hildebrandt M (2017) Saved by design? The case of legal protection by design. NanoEthics 11:307–311

Janseen H, Cobbe J, Singh J (2020) Personal information management systems: a user-centric privacy utopia? Internet Policy Rev 9(4):1–25

Janseen H, Cobbe J, Norval C, Singh J (2020) Decentralised data processing: personal data stores and the GDPR. Int Data Privacy Law 10(4):356–384

Koops B-J, Leenes R (2014) Privacy regulation cannot be hardcoded. A critical comment on the 'privacy by design' provision in data-protection law. Int Rev Law, Comput Technol 28(2):159–171

Larsen R, Brochot G, Lewis D, Eisma FL, Brunini J (2015) Personal data stores. European Commission—DG Connect. Report commissioned by the European Commission to the Cambridge University on personal data tools to evaluate feasibility and potential areas of policy assistance

Lehtiniemi T (2017) Personal data spaces: an intervention in surveillance capitalism? Surv Appl Math 15(5):626–639

Lu Y, Li S, Ioannou A, Tussyadiah I (2019) From data disclosure to privacy nudges: a privacy-aware and user-centric personal data management framework. In: Wang G, Bhuiyan MZA, De Capitani di Vimercati S, Ren Y (eds) Dependability in sensor, cloud, and big data systems and applications. Springer, Cham, pp 262–276

Mantelero A (2014) Social control, transparency, and participation in the big data world. J Internet Law 17(10):23–29

Poikola A, Kuikkaniemi K, Honko H (2014) MyData—a Nordic model for human-centered personal data management and processing

Solove DJ (2013) Introduction: privacy self-management and the consent dilemma. Harv Law Rev 126:1880–1903

Spagnuelo D, Ferreira A, Lenzini G (2020) Transparency enhancing tools and the GDPR: do they match? In: Mori P, Furnell S, Camp O (eds) Information systems security and privacy. Springer, Cham, pp 162–185

The Royal Society (2019) Protecting privacy in practice—the current use, development and limits of privacy enhancing technologies in data analysis

Zuboff S (2019) The age of surveillance capitalism: the fight for a human future at the new frontier of power. Public Affairs, New York

Explainability Due Process: Legal Guidelines for AI-Based Business Decisions

Camilla Tabarrini

1 Introduction

The Council of Europe defines Artificial Intelligence (AI) as the set of techniques aimed at reproducing human cognitive abilities through machine applications. The term AI was coined in 1956 during a Dartmouth College workshop, even though initial research on the subject dates back to before World War II. This notwithstanding, the fluctuating interest in AI applications skyrocketed only in the early 1990s due to the newest operative scenarios opened by the unprecedented data flow backing AI systems. Indeed, it was only at the dawn of the Big Data economy[1] that AI was finally able to unveil its true potential. More specifically, it was at the intersection between Big Data and Artificial Intelligence that originally stemmed the current emphasis on deep learning applications for decision-making purposes.

Nowadays, the use of AI to extract knowledge from large datasets of raw Big Data gives enterprises crucial competitive edge, irrespective of their size and line of business. Take incumbent Banks and FinTech start-ups, for example. Despite the highly regulated and traditional physiognomy of the banking and financial services market, incumbents were forced to orient their managerial styles and product design towards more consumer-centric and user-friendly standards. To this end, Big Data and AI play a pivotal role in enhancing the efficiency of customer experience and security and risk control, respectively, through the implementation of chatbots, robo-advice, customer segmentation as well as automated anomaly detection systems,

[1] For the purposes of this chapter Big Data economy and Industry 4.0 are used as synonyms referring to the heterogeneous class of business models based on the deployment of datasets characterized by the notorious five Vs: Volume, Variety, Velocity, Veracity and Variability (i.e. Big Data) to better orient their commercial strategies.

C. Tabarrini (✉)
Ca' Foscari University Venice, Venice, Italy
e-mail: camilla.tabarrini@unive.it

R. Senigaglia et al. (eds.), *Privacy and Data Protection in Software Services*, Services and Business Process Reengineering,
https://doi.org/10.1007/978-981-16-3049-1_18

payment transaction monitoring and cyber risk prevention. Furthermore, and more relevantly to the following legal analysis, AI technology allows for more accurate and comprehensive assessments of prospective clients' credit worthiness, thus fostering unbanked and underbanked financial inclusion, safer credit risk management and, in turn, financial stability. Beyond the banking sector, AI-based data analytics tools have seeped through every market of the so-called Industry 4.0 blowing the wind of customised advertising and user-centric marketing strategies.

Although delving into the data monetization implications of profiling exceeds the scope of this chapter, it is against the same AI background that originates the so-called 'black-box' problem underlying most AI-powered automated decision-making processes (Pasquale (2015); Citron and Pasquale (2014)). Indeed, moving from the Orwellian Chinese Social Credit System, passing through Western FICO-styled credit scoring applications, and landing to profiling for marketing purposes, the adoption of Big Data-based automated determinations raises the same explainability issues (Burrell (2016); Taylor (2017)).

The concept of AI explainability refers to the ability to trace back and translate into human-intelligible terms the algorithmic steps taken by the machine to produce an automated output. In other words, an AI decision-making process is explainable if both the programmer and the human recipient can understand why the machine reached a certain decision. This peculiar transparency threshold is generally within reach in the case of weak AI systems, but it becomes more challenging when it comes to strong AI applications. This triggers a general trade-off between accuracy and explainability of AI solutions. For instance, while Decision Trees and Linear and Logistic Regression represent a relatively interpretable but not particularly accurate model of Machine Learning Algorithm, Deep Learning subsets show higher performance standards at the cost of greater opacity.

These technological obstacles to a full disclosure of the algorithmic rationale, especially when dealing with high-stakes decisions, can undermine bias and error-detection auditing thus weakening societal acceptance of AI application for decision-making purposes. Hence, the objective of the chapter is to offer a brief overarching legal guideline on data protection requirements set by the General Data Protection Regulation (GDPR) in dealing with automated decisions based on personal data, addressing the issue of algorithmic transparency through controllers' information duties.

2 Article 22(1) and (2) GDPR: Scope of Application and Exceptions

Building on the similar provision enshrined in Article 15 of the former Directive 95/46/EC, Article 22(1) GDPR sets the right of any data subject "not to be subject to a decision based solely on automated processing, including profiling, which produces legal effects concerning him or her or similarly significantly affects him or her". Accordingly, for this Article to apply several requirements must be met.

First, the determination reached by the controller should be based *solely* on automated processing. As clarified by many legal scholars, this means that the decision is taken by the machine without any significant human involvement. Hence, the mere adoption of a 'human-in-the-loop' approach will not suffice. An alleged partially automated decision will fall outside the scope of application of Article 22 GDPR only if it is subject to human reviews carried out by someone with the knowledge and the power necessary to detect and correct eventual inaccuracies or simply change the final automated output due to a different evaluation of the input information. Moreover, contrary to the previous version of the provision, the automated data processing regulated by Article 22 *includes* and is not limited to profiling (Mendoza and Bygrave 2017). Hence, it applies not only to automated processing of personal data aimed at evaluating personal aspects relating to a natural person (*i.e.,* profiling, as defined by Article 4 GDPR) but also to neutral machine-based outputs. Take, for example, an automated speeding tickets system. Even if it merely turns input information ('detected car speed') into an automated output ('ticket') applying the speed-limit rules set by the programmer and without involving any assessment of the personal characteristics of the driver, it still falls within the scope of application of Article 22 GDPR.

Secondly, the automated decision should have legal or similarly significant effects on the recipient. As for the legal effects, during the COVID-19 pandemic, for instance, many States such as Italy used algorithmic methods to measure the contagion rates used to determine the extent of civil liberties restrictions needed to keep the outbreak under control. Clearly, this kind of algorithmic decision-making process was based on anonymized data thus falling outside the scope of application of the GDPR, still, it well reflects the potential pervasiveness of automated decisions. This notwithstanding, according to Recital 71 GDPR, Article 22(1) GDPR applies also to automated decisions that do not affect data subjects' rights but only other non-trivial aspects of their lives, such as eligibility to access financial, healthcare or even educational services. Therefore, no automated data processing activity is virtually exempt, including targeted advertising practices. Indeed, should it result in discriminatory customer segmentation, unfair commercial redlining, or even undue political influence over vulnerable internet users (especially when minors), targeted advertising could easily reach the threshold set by Article 22 GDPR.

However, despite its ambiguous wording, Article 22(1) GDPR is generally regarded as setting forth a passive data subjects' right entailing a general prohibition for any automated decision-making process presenting the above-described features. This means that controllers shall refrain from such activities irrespective of whether data subjects have actively invoked the right therein enshrined, unless such automated decision is based upon one of the exceptions set in the second paragraph.

More specifically, the controller is allowed to take a solely automated decision if it is based on the data subjects' explicit consent, it is necessary to enter into or perform a contract, or it is authorised by Union or Member State law.

As for the first exception, it is sufficient to stress that consent cannot be considered freely given (and therefore valid) if there exists a clear imbalance between data subjects and controllers. As exemplified in Recital 43 GDPR, this could be the case where the controller is a public authority. However, this presumption could be rebutted by showing that data subjects can withdraw their consent at any time without suffering any detriment. For instance, if an employer asks for employees' consent to process their personal data to offer them insurance coverage but denial has no unrelated repercussions, such consent can be deemed valid despite the power imbalance between the parties.

Secondly, automated data processing could be necessary to establish or carry out contractual obligations only if the controller can prove the lack of any realistic less intrusive alternatives. Therefore, a necessary automated decision, although not always indispensable, can never be merely useful in terms of enhanced efficiency or accuracy. Accordingly, pursuant to Article 22(2)(a) GDPR an automated decision is not necessary if used to improve a service, to introduce new functionalities or to exercise judicial or out-of-court rights pertaining to a pathological aspect of the contractual relationship (e.g., debt collection activities or judicial proceedings). Also, behavioural advertising cannot be deemed necessary to obtain an otherwise free online service just because used to finance it.

Finally, a solely automated decision is authorized by law only if such data processing technique is set out as compulsory by a clear and specific legal provision. On the contrary, if the controller is granted a margin of appreciation in the *an* and *quomodo* of the compliance, the automated data processing implemented by the controller using such discretion does not fall within the exception set forth in Article 22(2)(b) GDPR. For instance, this could be the case where a Bank decides to carry out an automated credit scoring process to comply with its creditworthiness-assessment duties generically set by Union and State law. Indeed, these regulations do not set a standard compulsory technique but merely require lenders to carefully and accurately assess the ability of a potential client to repay the loan he or she applied for. Hence, should a Bank decide to implement a fully automated credit scoring system because of its cost-cutting and accurate results, it cannot apply the law authorization exception to carry it out.

3 Article 22(3) GDPR: Suitable Safeguards and the Right to a Two-Phase Explanation

Pursuant to Article 22(3) GDPR if a solely automated decision meets all the conditions set out in the first paragraph and the controller is allowed to take such AI-based determination because of data subjects' explicit consent or contractual necessity, the former shall nonetheless implement suitable measures to safeguard data subject's rights. These safeguards, according to Recital 71 GDPR, should at least encompass data subjects' right to obtain specific information and a human intervention on the part of the controller, as well as to express their point of view, to obtain an explanation of the decision reached after such assessment and to challenge it.

According to the former Article 29 Working Party (hereinafter WP29, see Article 29 Working Party (2017)) an effective way to implement these safeguards would be to provide data subjects with on-request access to internal appeal procedures carried out by human agents after the automated decision has been reached. Moreover, in its guidelines on automated individual decision-making and profiling the WP29 expressly qualified these measures as a non-exhaustive list of good practice suggestions. Accordingly, the WP29 exemplified several complementary safeguards such as the implementation of regular algorithmic auditing processes to be carried out in-house or by third-party organizations. These quality checks should enable controllers to assess and prove whether the AI-based decision-making systems are working as intended, thus preventing erroneous or discriminatory results, and facilitating controllers' compliance with the accountability principle set by Article 24(1) GDPR.

This regulatory framework is further integrated by the controllers' information duties enshrined in Articles 13, 14 and 15 GDPR. More specifically, it is possible to distinguish between *ex ante* (i.e., *pre*-claim) and *ex post* (*i.e., post*-claim) information duties.

With regard to the *pre*-claim information duties, Articles 13(2)(f) and 14(2)(g) GDPR provide that, at the time personal data are collected (hence before any complaint on the automated output has been raised), controllers should inform data subjects about, among other things, "the existence of automated decision-making, including profiling, referred to in Article 22(1) and (4) and, at least in those cases, meaningful information about the logic involved, as well as the significance and the envisaged consequences of such processing for the data subject".

On the other hand, once such data processing activity has been initiated and for the entire time it is being carried out, data subjects have the right to obtain from the controller similar information (thus raising a claim). Indeed, despite the identical wording, most legal scholars have interpreted the information duty enshrined in Article 15(1)(g) GDPR as setting forth a right to an *ex post* (i.e.,*post*-claim) explanation of the automated decision reached by the controller.

Goodman and Flexman (2017) where the first to ever theorize a right to a human-interpretable explanation of algorithmic decisions stemming from the meaningful information about the logic involved required by Articles 13(2)(f), 14(2)(g) and 15(1)(g) GDPR, if read in conjunction with the suitable measures regulated by Article 22(3) GDPR and, more specifically, with the data subject's right to obtain an explanation of the decision reached as exemplified in Recital 71 GDPR.

The subsequent legal debate accordingly shifted on the content of such an explanation and the differences it may show in *pre*-claim and *post*-claim scenarios. Namely, the first interpretative dilemma pertained to the meaningfulness threshold of the information given set by Articles 13(2)(f), 14(2)(g), 15(1)(g) and 22 GDPR (Malgieri and Comandé (2017); Selbst and Powles (2017)).

In fact, as also recently stressed by the National Institute of Standards and Technology (NIST) of the U.S. Department of Commerce (2020), the second of the four principles of explainable Artificial Intelligence is the Meaningful principle. According to the NIST, this principle is met when the recipient is able to understand the explanation he or she has been given and, to this end, it also allows for tailored explanations at the level of both groups and individuals.

Along the same lines, while Article 12 GDPR clarifies that the information referred to in Articles 13, 14, 15 and 22 GDPR should be provided in a concise, transparent, and intelligible form, using clear and plain language, Article 5(1)(a) GDPR expressly calibrates this transparency standard with respect to the data subject. Therefore, for an explanation to be meaningful it must be tailored to the specific understanding of its individual recipient or the group of recipients the former belongs to.

Secondly, as further clarified by Recital 63 GDPR, every data subject should have the right to know the logic involved in any automatic processing of his or her personal data and, at least in the case of profiling, its consequences. Similarly, Article 9(1)(c) of the Modernised Convention for the Protection of Individuals with regard to the Processing of Personal Data of the Council of Europe (Convention 108+) recognizes the right of every data subject to obtain, on request, knowledge of the reasoning underlying automated decisions applied to him or her.

Furthermore, according to the explanatory report to the Convention 108+ such provision entails the data subjects' right to challenge the automated decision by putting forward their concerns regarding possible inaccuracies of the personal data used, as well as the irrelevance of the profile applied to them or of any other factor with an impact on the algorithmic result. Hence, the second feature of a meaningful explanation is for it to enable its recipients to detect data inaccuracies or irrelevant inferences before their use (*pre*-claim scenario), as well as to contest the automated output once it has been reached and applied to them (*post*-claim scenario).

Consequently, from this articulated regulatory framework stems a two-phase explanatory duty: (i) a necessary *pre*-claim description of the general features underpinning the algorithmic data processing method put in place by the controller and (ii) an eventual *post*-claim explanation of the specific automated decision reached to be provided on the recipient's request. Such two-pronged explanatory perspective is further corroborated by Article 12 GDPR. Indeed, on the one hand, Article 12(7) GDPR states that the information referred to in Articles 13 and 14 GDPR (i.e. the *pre*-claim explanation) should be aimed at providing data subjects with a meaningful overview of the intended processing, thus implying that the objective of a *pre*-claim explanation is to enhance data subjects' self-determination, by enabling them to predict the influence their actions might have on the automated decision-making process. On the other hand, Article 12(2) GDPR, if read in conjunction with the *post*-claim information duties enshrined in Articles 15(1)(g) and 22(3) GDPR, clarifies that this second explanatory step is mainly aimed at facilitating data subjects' right to contest the automated decision after it has been reached.

With this chronological and teleological distinction in mind, it is now possible to offer a more contextualized description of the operative content of each of these two explanations.

4 *Pre*-Claim Explanatory Duties

As already mentioned, to ensure that the automated decision-making system implemented is compliant with the fairness and transparency standards set by Articles 5(1) and 12(2) GDPR, controllers should make sure that, before any of their personal data are collected and used to take solely automated decisions, data subjects are being provided with meaningful information about the logic, significance and envisaged consequences of such data processing pursuant to Articles 13(2)(f) and 14(2)(g) GDPR.

In trying to translate this generic information duty into more operative terms little guidance is offered by the GDPR itself. However, helpful interpretative insights can be derived from the newest information duties introduced by the so-called "New Deal for Consumers" regulatory package. Namely, while Directive 2019/2161/EU amended Directive 2011/83/EU inserting a new Article 6a strengthening transparency in offers' rankings criteria, Regulation 2019/1150/EU introduced similar information duties in Platform to Business (P2B) relations as well. Indeed, both these regulations share with the GDPR the same objective: ensure adequate transparency in asymmetrical markets, especially when it comes to the rationale underpinning algorithmic decision-making mechanisms.

More specifically, pursuant to the amended Article 6a Dir. 2011/83/EU before a consumer is bound by any distant contract or similar offer, the online marketplace provider shall provide general information on the main parameters determining ranking of offers presented to the consumer as a result of the search query, as well as the relative importance of those parameters as opposed to other parameters. Similarly, according to Article 5 Reg. 2019/1150/EU providers of online intermediation services and providers of online search engines shall provide business users with a publicly available description of the main parameters determining ranking and the reasons for the relative importance of those main parameters as opposed to other parameters.

Just like the solely automated decisions regulated by Article 22 GDPR can significantly affect data subjects' rights, the use of algorithmic sequencing, rating, or ranking often has decisive influence over the commercial lives of business users therefore exacerbating the call for non-arbitrary and predictable automated saliency tools.

To this end, Recital 24 Reg. 2019/1150/EU and Recital 22 Dir. 2019/2161/EU clarify that algorithmic predictability entails a concise and intelligible outline of the main parameters incorporated into an automated ranking mechanism. This means that online search engines and marketplace providers should identify a limited set of the most significant ranking criteria out of a possibly much larger number of parameters affecting the result. Furthermore, according to Recital 25 Reg. 2019/1150/EU such a reasoned description of the main ranking parameters should also include an explanation of how business users can actively influence the ranking, for example, by describing which features of the goods and services offered are taken into account by the automated saliency tool. In this way business users would be enabled to better foresee the impact a product design variation might have on their platform visibility, thus also allowing them to compare and choose the platform using the ranking system that better suits their products.

Such augmented regulatory maturity and detail can be particularly helpful to shed an interpretative light on the informative duties set by the GDPR. Indeed, the mentioned rationale similarities between these two sets of regulations make it possible to infer that what the GDPR regulator meant by meaningful overview of the intended processing (as referred to in Recital 60 GDPR) was a reasoned description of the main data processing criteria used by the AI-system to reach the automated decision. Accordingly, pursuant to Articles 13(2)(f) and 14(2)(g) GDPR disclosing meaningful information about the logic involved in an automated decision-making process means providing data subjects with the most significant algorithmic steps taken by the machine to infer the knowledge used to reach the determination. In other words, the *pre*-claim description of the algorithmic logic followed by the machine can be deemed meaningful pursuant to the GDPR if, out of all the parameters potentially affecting the automated decision, controllers disclose only those most suitable to enable data subjects to understand how they can influence the algorithmic outcome, thus changing their behaviour accordingly. Therefore, in the *pre*-claim scenario the meaningfulness threshold conveys a predictability standard tied to a more effective exercise of data subjects' self-determination.

Also, from an IP standpoint, and reproducing the wording of Recital 63 GDPR, Recital 27 Reg. 2019/1150/EU clarifies that compliance with Dir. 2016/943/EU should not result in a refusal to disclose even the main ranking parameters. However, along the lines of its mentioned enhanced regulatory maturity, Recital 27 goes further and stresses how, irrespective of any trade secret protection limitations, the reasoned description "should at least be based on *actual* data on the relevance of the ranking parameters used".

This last provision takes another step closer to operative clarity. Indeed, it helps overcoming the interpretative impasse concerning the alternative between a model-centric and a subject-centric explanation.

Indeed, as well exemplified by Edwards and Veale (2017), a model-centric explanation consists of an abstract description of the general functioning of an AI system, while a subject-centric standard refers to a variegate plethora of context-based explanatory techniques sharing a stronger focus on the specific circumstances of the data processing carried out. In other words, a subject-centric explanation is always aimed at disclosing the rationale behind an individual decision as opposed to the general methodological overview offered through a model-centric approach. For this reason, given the generic and hypothetical nature of a model-centric explanation, only a subject-centric explanatory technique could be based on the actual data of the case and not only on synthetic data.

In this context, the reference made by Recital 27 Reg. 2019/1150/EU to the need to base the explanation on *actual* data on ranking criteria's relevance appears to orient the explanatory design process towards subject-centric models fed by a case-specific data flow. This means that a meaningful explanation of an automated saliency tool entails a non-standardized (*i.e.*, non-model centric) description of the main ranking parameters used selected among those more relevant according to the specific circumstances of the case. Accordingly, controllers should design multi-faced explanations that, although not necessarily individualized, take into account the different relevance each variable may show towards different groups of customer-segmented recipients.

Such interpretative assumptions are further corroborated by Recital 15 Reg. 2019/1150/EU where it stresses that unspecific and generic terms and conditions shall not be deemed intelligible inasmuch as they fail to give business users a reasonable degree of predictability on the most important aspects of the contractual relationship.

In conclusion, a *pre*-claim explanation of a solely automated decision pursuant to Articles 13, 14 and 22 GDPR should be group-tailored and could resemble the following statement: the car insurance rate is determined on the basis of multiple variables such as drivers' age and employment, vehicle type, crash reports and residence zip code. For Y-to-Z aged drivers the most relevant factor is generally age itself, weighing on the K% of the final rate, followed by occupation at J% and zip code at I%. On the other hand, for drivers aged X-to-Y the most relevant parameters are occupation and zip codes both weighing on the V% of the final rate. However, if insured drivers live in AAAA zip code area, the insurance rate will generally be subject to a T% increase irrespective of their age (Fig. 1).

In addition, for this explanation to allow recipients to evaluate the fairness of the exemplified automated insurance-rating system, it should be integrated with a brief description of the reasons behind the different weight each parameter may have in relation to different drivers' sub-categories. This explainability threshold could be also reached through an interactive and multi-layered interface where users

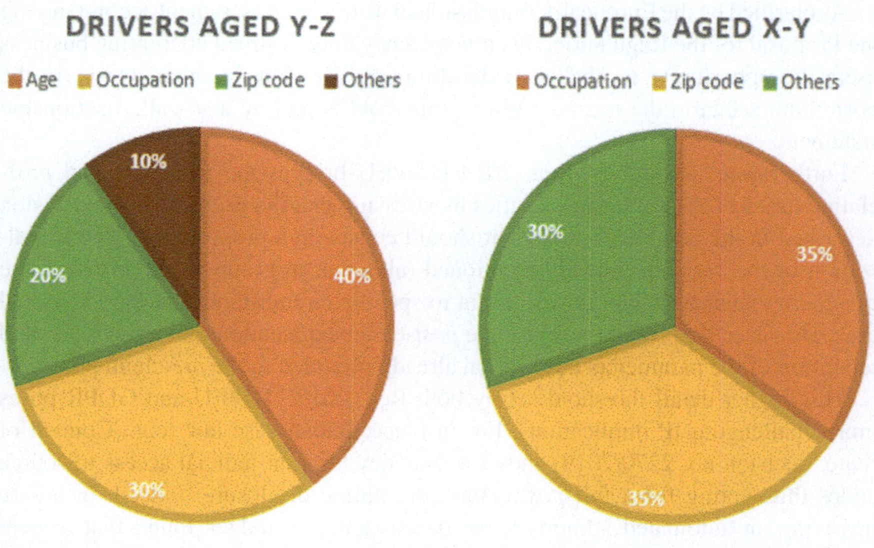

Fig. 1 The image visually represents the relative weight of each variable mentioned in the explanation thus mirroring the different balances reached with regard to different groups of clients

could insert their personal data and be guided towards their best-fitted *pre*-claim group-based explanation.

5 *Post*-Claim Explanatory Duties

According to Articles 12(1), 15(1)(h) and 22(3) GDPR, controllers should design the whole decision-making process in a way that allows data subjects to obtain human intervention on the part of the controller and to express their opinions. Also, once an automated decision has been reached, controllers should provide data subjects with all the information necessary to not only enable but also facilitate the exercise of the data subjects' right to contest the automated decision applied to them. Hence, the last interpretative step is to determine which kind of algorithmic disclosure can amount to a meaningful *post*-claim explanation pursuant to the GDPR.

To this end, Reg. 2019/1150/EU once again offers helpful interpretative insights. Indeed, at Recital 24 it states that when online intermediation service providers decide to take any demoting action negatively affecting a business user's appearance on their platform (also known as 'dimming'), they should provide the recipient with a statement of reasons for such a decision and implement an internal complaint-handling process. The Recital also adds a teleological clarification by stressing that such a statement of reasons should be drafted in a way that allows business users "to ascertain whether there is scope to challenge the decision, thereby improving the possibilities for business users to seek effective redress where necessary".

As clarified by the European Commission in its Impact Assessment accompanying the Proposal for the Regulation, this transparency duty is aimed at granting business users the opportunity to challenge the dimming decision. In other words, in the *post*-claim scenario the meaningfulness threshold is met by a so-called actionable statement.

Furthermore, Recital 24 Reg. 2019/1150/EU bridges the *pre*-claim and *post*-claim phases of the explanatory duties by stressing that the grounds of the decision described in the actionable statement should consist in a proportionate contextualization of the variables already mentioned (although in group-based terms) in the *pre*-claim explanation and now relevant to specific circumstances of the individual case. However, this should not turn the *post*-claim explanation into a mere detailed repetition of the parameters description already provided in the *pre*-claim phase.

This higher detail threshold set by both Reg. 2019/1150/EU and GDPR poses more challenging IP implications. For instance, Italian case law (e.g., Council of State, decision no. 2270/2019) shows a tendency to grant judicial access to source codes thus opting for a full *post*-claim algorithmic disclosure. Indeed, in a case involving an automated administrative decision Italian judges found that a mere individual access to the source code could not impair controllers' copyrights over it since Dir. 2016/943/EU does not prevent *any* access but only those potentially able to undermine the financial exploitation of the decision-making software.

However, European regulations clearly suggest otherwise. Indeed, according to Recital 63 GDPR, data subjects' right to obtain knowledge of the logic involved in any automatic personal data processing should not infringe upon the IP limitations set by both Dir. 2009/24/EC on the legal protection of computer programs and Dir. 2016/943/EU on the protection of trade secrets. Similarly, Recital 23 Dir. 2019/2161/EU stresses that the reasoned description of the main ranking parameters should not interfere with the trade secret protection and, to this end, should not involve any disclosure of the algorithms underpinning the saliency tools. The same exclusion is conveyed to the P2B context by Recital 27 Reg. 2019/1150/EU where it is also pointed out how a more general criteria description satisfies users' transparency needs while also preventing third-party bad faith manipulations of ranking mechanisms.

Accordingly, despite the ongoing legal debate concerning the extent of a meaningful *post*-claim disclosure, it is possible to affirm that, at least at the internal complaint-handling level, an actionable statement should not necessarily entail complainants' access to the algorithm or the source code. In fact, such explanatory approach would be neither intelligible (*i.e.,* meaningful) nor IP proportionate since it would most probably trigger lengthy judicial proceedings and, in any case, force data subjects to bear the costs of IT experts consultations to translate complex algorithmic formulas into actionable statements within the meaning above-described.

At the same time, however, post-*claim* information duties could neither be downgraded to the mere disclosure of the one criterion conveying the smallest change the data subject should make to obtain the desired outcome as suggested by Wachter et al. (2017) in theorizing the idea of a so-called "counterfactual explanation". This interpretative approach aims at providing a data subject with a description of the variable that, no matter its weight on the outcome, represents the smallest change the recipient should make to her/his input data to obtain the desired outcome (so-called 'closest possible world'). This parameter, however, will intuitively be far from the most significant to unveil the inner logic followed by the machine or to ascertain whether there are grounds to contest it. Indeed, this counterfactual explanation would not allow data subjects to know which of all the variables used by the AI system to reach the automated decision weighted the most in their specific case, but only the one the recipient could most easily use to change the algorithmic output.

In light of the above, the most useful but still IP-sensitive way to comply with the *post*-claim information duties enshrined in Articles 15(1)(h) and 22(3) GDPR is to disclose the profile inferred from the raw personal data and applied to a single recipient by association. Although this piece of information could still potentially amount to a trade secret, it conveys all the knowledge the recipient may need to assess the accuracy and the fairness of the automated decision. Hence, a meaningful *post*-claim explanation could resemble the following statement: "we have analyzed the following categories of your personal data (…) in combination with a pool of raw Big Data obtained from these internal and third-party sources (…). We have used a deep/weak AI supervised/unsupervised decision-making system through which it has been inferred that people sharing with you these features (…) present these characteristics (…). Accordingly, our automated decision is that (…)".

Consequently, it is hereby stressed that controllers should make sure that any AI system they choose to adopt to take high-stake automated decisions within the described meaning of Article 22 GDPR should allow them to trace back all these algorithmic steps; otherwise they won't be able to comply with their explainability duties pursuant to the GDPR.

References

Article 29 Working Party (2017) Guidelines on Automated individual decision making and profiling for the purposes of Regulation 2016/679. https://ec.europa.eu/newsroom/article29/item-detail.cfm?item_id=612053. Accessed 30 Dec 2020

Burrell J (2016) How the machine 'thinks': understanding opacity in machine learning algorithms. Big Data & Society 3(1)

Citron DK, Pasquale F (2014) The scored society: due process for automated predictions. Washington Law Rev 89:1–10

Edwards L, Veale M (2017) Slave to the algorithm? Why a 'Right to an Explanation' is probably not the remedy you are looking for. Duke Law Technol Rev 16(1):18–84

Goodman B, Flaxman S (2017) European Union regulations on algorithmic decision-making and a "right to explanation". AI Magazine 38(3)

Malgieri G, Comandè G (2017) Why a right to legibility of automated decision-making exists in the general data protection regulation. Int Data Privacy Law 7(4):243–265

Mendoza I, Bygrave LA (2017) The right not to be subject to automated decisions based on profiling. In: Synodinou TE, et al (eds) EU Internet Law. Regulation and Enforcement. Springer, Heidelberg, p 79

National Institute of Standards and Technology (2020) Four principles of explainable artificial intelligence. https://doi.org/10.6028/NIST.IR.8312-draft

Pasquale F (2015) The black box society. Harvard University Press, Cambridge, The secret algorithms that control money and information

Selbst AD, Powles J (2017) Meaningful information and the right to explanation. Int Data Privacy Law 7(4):233–242

Taylor R (2017) No privacy without transparency. In: Leenes R et al (eds) Data protection and privacy: the age of intelligent machines. Hart Publishing, Oxford, p 77

Wachter S, Mittelstadt B, Floridi L (2017) Why a right to explanation of automated decision-making does not exist in the general data protection regulation. Int Data Privacy Law 7(2):76–99

The manufacturer's authorised representative in the EU is Springer Nature Customer Service Centre GmbH, Europaplatz 3, 69115 Heidelberg, Germany. If you have any concerns regarding our products, please contact ProductSafety@springernature.com

Printed and bound by CPI Group (UK) Ltd, Croydon, CR0 4YY
30/09/2026
02225295-0003